Regulatory Compliance Monitoring
By
Atomic Absorption Spectroscopy

Regulatory Compliance Monitoring By Atomic Absorption Spectroscopy

By
Sidney A. Katz
and
Stephen W. Jenniss

Verlag Chemie International
Deerfield Beach, Florida

Sidney A. Katz, Ph.D.
Associate Dean
Camden College of Arts and Sciences
Rutgers - The State University of New Jersey
Camden, New Jersey 08102

Stephen W. Jenniss
Quality Assurance Coordinator
Supervisor of the Office of Quality Assurance
Division of Water Resources
New Jersey Department of Environmental Protection
Trenton, New Jersey 08625

Library of Congress Cataloging in Publication Data

Katz, Sidney A., 1935-
  Regulatory compliance monitoring by atomic absorption spectroscopy.

      Includes bibliographical references and index.
      1. Atomic absorption spectroscopy. 2. Environmental chemistry. I. Jenniss, Stephen
W., 1952-
II. Title. III. Series.
QD96.A8K37 1983   543'.0858   83-14719
ISBN 0-89573-114-2

ISBN 0-89573-114-2   Verlag Chemie International, Deerfield Beach
ISBN 3-527-26058-7   Verlag Chemie GmbH, Weinheim

Preface

The preparation of this manual was undertaken at the suggestion of the publisher following our presentation of a paper on the applications of atomic absorption spectrometry to environmental monitoring at the 1980 Eastern Analytical Symposium. We felt, at that time, that there was a need to collect the regulatory requirements into a single volume for those using atomic absorption spectrometry in compliance monitoring. We have deliberately avoided a rigorous and detailed presentation of the theory and instrumentation for atomic absorption spectrometry, assuming that those employing this technique are already familiar with these aspects. We have, instead, concentrated on the applications of atomic absorption spectrometry to regulatory compliance monitoring.

This manual is directed to both the procedures preceding the absorbance measurements and the conditions under which atomic absorbance is measured. Sample collection, sample preservation, and sample dissolution are considered in the former category, and flame/furnace parameters, wavelength selections, interferences, and quantation are included in the latter. We have, in addition, included a chapter on quality control/quality assurance. For the most part, we have relied upon the published requirements and recommendations of the regulatory agencies in preparing this manual.

We are grateful to the Environmental Protection Agency (USA), the Occupational Safety and Health Administration (USA), the Department of the Environment (Canada), the Institute for Water, Soil and Air Hygiene of the Federal Health Department (Federal Republic of Germany), and the Department of the Environment (UK) for providing us with their respective published procedures on the use of atomic absorption spectrometry in compliance monitoring. Our respective institutions are acknowledged for their encouragement in undertaking this work. Special thanks go to Doctor Edmund Immergut, consulting editor to the publisher, for his initial suggestion of the topic and his continual guidance in the course of preparing the manuscript. We are most grateful to Olga Moore for the cover design and to James Hammond and Kathleen Giordano for Figures 3.1, 8.1, and 8.2. Expressions of appreciation are insufficient recognition of the time and effort Ms. Theresa Watson spend at the typewriter transcribing and revising the text. Finally, our respective wives, Sheila and Shelly, deserve special mention for their patience and forebearance during the completion of this work.

Contents

1. INTRODUCTION

In as short a time as 15 years, atomic absorption spectrometry has become the most commonly used technique for determining trace elements, usually metals, in a wide variety of environmental and biological materials. This rapid growth is due not only to the sensitivity and selectivity of atomic absorption spectrometry, but also to its speed, simplicity, and broad scope.

1.1. Scope and Sensitivity of Atomic Absorption Spectrometry

Some 70 elements can be determined by atomic absorption spectrometry. For most of these, detection limits, according to manufacturers' literature,[1-3] are in the range of from a few parts per billion using electrothermal atomization techniques to a few parts per million when conventional flame atomization techniques are employed. Detection limits are frequently two to three decades lower with electrothermal atomization and/or hydride generation techniques than with conventional flame atomization techniques. Welz[4] has summarized the experimental conditions and detection limits for six dozen elements, and Van Loon[5] has concisely presented over 150 procedures for the determination of trace elements in a wide variety of materials by atomic absorption spectrometry. The experimental conditions[6,7] and the optimum ranges[8,9] for the determination of some two dozen elements of environmental and biological consequence are tabulated in Table 1.1, and the regulatory limits on some trace element levels in air and water are summarized in Table 1.2.

1.2. Theory of Atomic Absorption Spectrometry

Atomic absorption spectrometry is based upon the absorption of resonance radiation by an atomic vapor of the analyte. The resonance radiation corresponds to the wavelengths

1

Table 1.1. Parameters Associated with the Determination
of Some Frequently Encountered Elements by
Atomic Absorption Spectrometry

| | | | | Flame technique | | |
Element	Wavelength	Fuel	Oxidant	Flame conditions	Additive
Ag	328.1	C_2H_2	air	Lean-blue	—
Al	309.2	C_2H_2	N_2O	Rich-red	KCl
As	193.7	H_2	air-Ar	—	—
Ba	553.6	C_2H_2	N_2O	Rich-red	KCl de-ionizer
Be	234.9	C_2H_2	N_2O	Rich-red	—
Ca	422.7	C_2H_2	air	Rich-yellow	$LaCl_3$ re-leaser
Cd	228.8	C_2H_2	Air	Lean-blue	—
Co	204.7	C_2H_2	Air	Stoichiomet	—
Cr	357.9	C_2H_2	Air	Sli. rich	—
Cu	324.7	C_2H_2	Air	Lean-blue	—
Fe	248.3	C_2H_2	Air	Lean-blue	—
Hg	253.7	—	—	—	—
K	766.5	C_2H_2	Air	Sli. lean	—
Mg	285.2	C_2H_2	Air	Rich-yellow	$LaCl_3$ re-leaser
Mn	279.5	C_2H_2	Air	Lean-blue	—
Mo	313.3	C_2H_2	N O	Rich-red	$Al (NO_3)_3$
Na	589.6	C_2H_2	Air	Lean-blue	—
Ni	232.0	C_2H_2	Air	Lean-blue	—
Pb	283.3	C_2H_2	Air	Sli. lean	—
Sb	217.6	C_2H_2	Air	Lean-blue	—
Se	196.0	H_2	Air-Ar	—	—
Sn	286.3	C_2H_2	Air	Rich-yellow	—
Ti	365.3	C_2H_2	N_2O	Rich-red	KCl
Tl	276.8	C_2H_2	Air	Lean-blue	—
V	318.4	C_2H_2	N_2O	Rich-red	$Al (NO_3)_3$
Zn	213.9	C_2H_2	Air	Lean-blue	—

^{a} ppb.

Working range (ppm)	Special notes	Furnace technique		Working range (ppb)
		Drying, ashing, and atomizing time (s) at temperature (°C)		
0.1-4	—	30/125, 30/400, 10/2700		1-25
5-100	—	30/125, 30/1300, 10/2700		20-200
2-20[a]	Hydride generator	30/125, 30/1100, 10/2700		5-100
1-20	—	30/125, 30/1200, 10/2800		10-200
0.1-2	—	30/125, 30/1000/ 10/2800		1-30
0.1-20	—	—		—
0.1-2	—	30/125, 30/500, 10/1900		0.5-10
0.5-10	—	30/125, 30/900, 10/2700		5-100
0.2-10	—	30/125, 30/1000, 10/2700		5-100
0.1-10	—	30/125, 30/900, 10/2700		5-100
0.3-10	—	30/125, 30/1000, 10/2700		5-100
2-20[a]	cold vapor	—		—
0.1-2	—	—		—
0.1-2	—	—		—
0.1-10	—	—		—
5-20	—	30/125, 30/1000, 10/2700		1-30
0.1-1	—	30/125, 30/1400, 15/2800		3-60
0.5-10	—	—		—
1-20	—	30/125, 30/900, 10/2700		5-100
1-40	—	30/125, 30/500, 10/2700		5-100
2-20[a]	Hydride generator	30/125, 30/800, 10/2700		20-300
10-200	—	—		—
5-100	—	30/125, 30/600, 10/2700		20-300
1-20	—	30/125, 30/1400, 15/2800		50-500
1-100	—	30/125, 30/1400, 15/2800		10-200
0.1-2	—	30/125, 30/400, 10/2500		0.2-4

Table 1.2. Regulatory Limits on Trace Element
Levels in Air and Water

Element	Airborne limit (mg/m^3)	Drinking water limit (mg/L)
Arsenic		0.05
Inorganic	10	
Organic	0.5	
Barium	0.5	1
Beryllium	0.002	
Boron	15	
Cadmium		0.01
Dust	0.2	
Fumes	0.1	
Calcium	5	
Chromium		
Hexaval.	0.1	0.05
Insol.	1	
Sol.	0.5	
Cobalt	0.1	
Copper		1
Dust	1	
Fume	0.1	
Iron	10	0.3
Lead	0.05	0.05
Magnesium	15	
Manganese	5	0.05
Mercury	0.1	0.002
Nickel	1	
Selenium	0.2	0.1
Silver	0.01	0.05
Tin		
Inorganic	2	
Organic	0.1	
Vanadium		
Dust	0.5	
Fumes	0.1	
Zinc		5
Chloride	1	
Oxide	5	

associated with the excitation of gaseous analyte atoms from
their ground states to excited states. Most elements show
relatively simple, well-defined, characteristic absorption
spectra. There is little likelihood of spectral interference
when the excitation radiation is of high spectral purity.
Hence, atomic absorption spectrometry is highly selective.

When resonance radiation characteristic of a given element
is incident upon a cell containing that element as an atomic
vapor, absorption occurs in the direction of the incident
radiation. The excited atoms can and do return to the ground
state, but the resulting emissions do not compensate for the
absorption because the former is isotropic.

In a cylindrical cell of radius r and length 1 containing
C gaseous atoms per unit volume with N_o resonance radiation
photons entering per unit time per unit area, the rate of
photon absorption is:

$$-dN = NkC\pi r^2 dx / \pi r^2$$

where k is the reaction cross section for resonance absorption
and $C\pi r^2 dx$ is the number of gaseous atoms in the element of
volume under consideration.

$$-\int_{N_o}^{N} dN/N = kC \int_{o}^{\ell} dx$$

$$\ln N_o/N = kC\ell$$

The logarithm of the ratio of the incident photon flux to the
emergent photon flux is the absorbance of the resonance radiation. Because the volume of the cell is constant, the absorbance is directly proportional to the number of gaseous
atoms in the system. This proportionality is the basis for
atomic absorption measurements. Conditions must be controlled
to insure this proportionality is maintained. Atomic absorption spectrometry is, obviously, a comparative rather than an
absolute measurement technique.

1.3. Requirements for Atomic Absorption Spectrometry

The measurement of atomic absorbance requires a source of resonance radiation, a monochromator for isolating the resonance lines, a detector for determining the intensities of the incident and emergent photon fluxes, and an atomizer for generating the atomic vapor of analyte.

1.3.1. Sources of Resonance Radiation

Atomic spectral lines typically have natural widths on the order of 10^{-5} nm. Isolation of light with this degree of spectral purity from a continuum source demands a monochromator of very high resolution and slit widths so narrow that stimulation of the photoelectric detector by the feeble emergent beam is minimal. The hollow cathode lamp is a simple alternative to meeting these demands.

The hollow cathode lamp (HCL) consists of a glass cylinder, with an appropriate optical window, containing a cup-shaped cathode fabricated from the same element as contained in the analyte (Figure 1.1). The lamp is filled with argon or neon at a pressure of from 5 to 10 torr. At an applied potential of a few hundred volts and currents in the approximate range of 10 to 20 ma, atoms of the filling gas undergo ionization and bring about the sputtering and excitation of the cathode material within the hollow cup. The missions are essentially the line spectra of the cathode material and of the filling

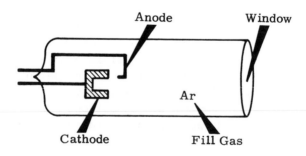

Figure 1.1. Hollow cathode lamp (HCL). Reproduced with permission from the Perkin-Elmer Corporation.

gas. The resonance absorption line of the analyte is certainly contained in the emission spectrum of the cathode material.

The control of lamp current is important. As the current is increased, both sputtering and excitation of the cathode material increase. The former gives rise to a high concentration of gaseous atoms from the cathode material within the lamp. These absorb the resonance radiation. Self-absorption broadening results in degradation of the sensitivity and curvature of the calibration line. Ideally, the current should be sufficient to produce an intense resonance line with minimal self-absorption broadening. Best guidance on lamp currents is obtained from the manufacturers' literature.

The electrodeless discharge lamp (EDL) overcomes some of the difficulties encountered with hollow cathode lamps for the more volatile elements whose resonance lines are far in the ultraviolet (Figure 1.2). Electrodeless discharge lamps for arsenic and selenium produce much more intense radiation than the corresponding HCLs. The useful life of a hollow cathode lamp operated at the current required to produce resonance radiation of intensity equal to that of an EDL is quite short. Electrodeless discharge lamps are not limited to arsenic and selenium. They are currently available for arsenic, bismuth, cadmium, cesium, germanium, mercury, phosphorus, lead, rubidium, antimony, selenium, tin, tellurium, thallium, and zinc. The operation of these lamps requires a special power supply.

Electrodeless discharge lamps produce resonance radiation by high-frequency excitation of the desired element. The excitation takes place in a small sealed quartz tube (1 cm ×

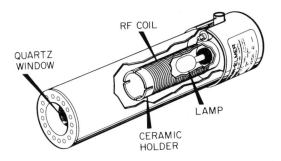

Figure 1.2. Electrodeless discharge lamp (EDL). Reproduced with permission from the Perkin-Elmer Corporation.

5 cm) or sphere (1.5 cm diameter) containing a few milligrams of the element or one of its compounds, usually the iodide. The quartz tube also contains argon, xenon, or krypton at a few torr. The quartz container is situated within the coil of a high-frequency generator.

The electrodeless discharge lamp requires some 30 min for stabilization. After this warmup period, the spectral output is some 10 to 100 times greater than that of a corresponding HCL. The light emitted is not subject to self-absorption.

1.3.2. Monochromators

The function of the monochromator is to isolate a single atomic resonance line from the HCL or EDL and to reject emissions from the atomizer. Ideally, the monochromator should be capable of isolating only the resonance line for the analyte and excluding all others. For some elements; ie, copper, this is readily accomplished; for others, ie, nickel, complete isolation is not achieved. The spectrum of copper is relatively uncluttered in the vicinity of the 324.8-nm resonance line. The nearest line in the copper spectrum is at 327.4 nm. The 232.0-nm nickel resonance line, on the other hand, is bounded with other lines at 231.7 nm and 232.1 nm. Isolation of the nickel resonance is accomplished, in part, by narrowing the slit. The former nonabsorbing line is particularly troublesome, and it is responsible for the nonlinearity of the calibration.

Most atomic absorption spectrometers employ gratings which are capable of spectral band passes at least as narrow as 0.1 nm. In addition, most atomic absorption spectrometers mechanically or electronically chop the output of the lamp. Such modulation allows electronic differentiation between the intensity of the resonance line and emissions from the atomizer.

1.3.3. Detectors

Essentially all commercially produced atomic absorption spectrometers employ the photomultiplier tube (PMT) as the detector. Resonance radiation photons from the HCL or EDL pass through the atomizer and the monochromator, and they enter the PMT through an appropriate optical window. These photons interact with the photocathode material and bring about the

ejection of photoelectrons. The photoelectrons are attracted
to the first dynode stage under the influence of a potential
gradient. The photoelectrons acquire sufficient kinetic
energy so that each causes the ejection of several secondary
electrons by interaction at the first dynode stage where the
process is repeated. Current amplifications of 10^6 or 10^7
are achieved in a 10 dynode stage PMT.

The photocathode material determines the PMTs spectral re-
sponsiveness and sensitivity. The cesium-antimony cathodes
demonstrate adequate sensitivity in the range of from 200 nm
to 700 nm. The trialkali-antimony (sodium, potassium, and
cesium) are of somewhat lower sensitivity, but they respond to
photons of energies corresponding to the spectral range of
200 nm to 800 nm.

1.3.4. Atomization Devices

The efficient and reproducible introduction of the analyte
as an atomic vapor is, perhaps, the most difficult aspect of
atomic absorption spectrometry. With the cold vapor technique
for mercury as the single possible exception, atomization is
achieved by thermal decomposition of solid, liquid, or gaseous
forms of the analyte. Both flames and furnaces are used for
this purpose.

1.3.4.1. Flames   The production of an atomic vapor from a
solution of the analyte by flame atomization involves five
staps: (1) nebulization, (2) droplet precipitation, (3) mixing, (4) desolvation, and (5) compound decomposition. The
sample solution is drawn through a capillary into the nebulizer
at a rate of between 3 and 6 mL/min by the venturi action of
the flowing oxidant. Some of the liquid, approximately 5%, is
dispersed as a fine aerosol mist. The larger droplets, most
of the aspirated material, precipitate and flow off to waste.
The aerosol mist is mixed with the fuel and oxidant gases and
emerges from the burner head into the flame. The aerosol
particles undergo desolvation in the flame. The resulting
solid particles are decomposed into their constituent atoms
by the flame, thereby creating the environment of gaseous
atoms necessary for atomic absorption.

Since atomic absorption spectrometry is a comparative
technique, it is necessary that atomization from the sample
solutions take place to the same extent as atomization from

the reference standards. Differences in viscosity between
the samples and the standards will affect the aspiration rate
and ultimately the number of gaseous analyte atoms. Differences in surface tension have a direct effect on nebulization
efficiency and, hence, affect the number of gaseous analyte
atoms. The anion of the metal compound analyte determines the
ease of decomposition into gaseous atoms. It is necessary,
therefore, to match the reference standards as closely as possible to the matrices of the sample solutions, or to modify
the sample matrices to match those of the standards.

Most atomic absorption spectrometers accept both a 10-cm
air-acetylene burner head and a 5-cm nitrous oxide-acetylene
burner head. The air-acetylene flame is essentially the
standard for atomic absorption spectrometry. At least three
dozen elements can be determined using this flame. It pro-
duces a temperature of 2300°C, and it shows good optical
properties down to 230 nm. The nitrous oxide-acetylene flame
reaches a temperature of 3000°C, making possible the determination of some of the more refractory elements as well as
eliminating some of the chemical matrix interferences. Its
optical properties are also good. Both the air-acetylene
flame and the nitrous oxide-acetylene flame absorb significantly below 200 nm. The air entrained argon-hydrogen flame
is more transparent at this wavelength. Hence, it is frequently used for the measurement of absorbances due to arsenic
and selenium, whose resonance lines are at 193.7 nm and
196.1 nm, respectively.

1.3.4.2. Hydride Generators   Arsenic, bismuth, germanium,
antimony, selenium, tin, and tellurium form gaseous hydrides
under easily controlled conditions. It is possible to atomize
these elements from their hydrides and thereby improve sensitivity and reduce interference in their respective determinations.

The hydride generator consists of: (1) a reaction vessel
in which the hydride is formed, (2) a means of adding the reducing agent to form the hydride, and (3) a means of intro-
ducing the gaseous hydride to the atomic absorption spectrometer. Each manufacturer of atomic absorption spectrometers
markets a hydride generator system that carries out these
functions. In most systems, the atomization is carried out in

a heated cell.* At least one generator system introduces the
hydride into the mixing chamber of the burner so that atomization can take place in the air entrained argon-hydrogen
flame.†

Determinations made on the basis of hydride atomization are
much more selective and sensitive than those made by aspirating
solutions. The formation of the gaseous hydride and its subsequent introduction into the atomic absorption spectrometer
leave behind any matrix interferences that the sample may
have. The fraction of analyte actually atomized approaches
100% when hydride generation techniques are employed. In addition, absorbance by the flame is eliminated when atomization
takes place in the heated tube. The improved sensitivity of
hydride generation techniques is summarized in Table 1.3.

1.3.4.3. Cold Vapor Generators Mercury is unique among
the elements amenable to atomic absorption spectrometry: elemental mercury has an appreciable vapor pressure at room temperature (20 mg/m^3 at 25°C). Hence, it can be swept into a
cell positioned in the optical path of an atomic absorption
spectrometer for measurement of absorbance.

The cold vapor generator consists of: (1) a reaction vessel in which the mercury is reduced to the elemental form,
(2) a means for adding the reducing agent, and (3) a means of
sweeping the atomic vapor into the cell. Many of the hydride
generating systems can also serve as cold vapor generators.
In such applications, the cell is not heated. Detection
limits for cold vapor generation are some five decades lower
than those obtained with conventional flame atomization, ie,
0.001 μg/L versus 170 μg/L.

1.3.4.4. Furnaces Furnaces are the most common of the
electrically heated devices used to produce atomic vapors.

---

*Varian Model 65 Vapor Generation Accessory, IL AVA (Instru-
    mentation Laboratory Atomic Vapor Accessory), Baird Arsenic
    Selenium Detection Unit, and Perkin-Elmer MHS-10 Mercury/
    Hydride System.

†Perkin-Elmer Hydride Generation Sampling System 303-0849.

Table 1.3. Comparison of Detection Limits
for Flame Atomization and Hydride
Generation Techniques

	Detection limits (µg/L)	
Element	Flame	Hydride
Antimony	30	0.1
Arsenic	140	0.02
Bismuth	20	0.02
Selenium	70	0.02
Tellurium	20	0.02
Tin	110	0.5

These devices are much more efficient than flames in producing atomic vapors. Hence, furnace techniques require only micro-liter amounts of sample, and they are frequently two to three decades more sensitive than flame techniques as reflected by comparison of the working ranges listed in Table 1.1.

Electrothermal atomization usually takes place in an inert atmosphere, and it involves three distinct, preprogrammed, time-temperature steps for drying, ashing, and finally atom-izing the sample. The conditions of time and temperature for each of these steps depend upon both the analyte and the sample.

The drying step serves to remove solvent and to deposit the sample as a finely divided solid film on the floor of the fur-nace. This is usually accomplished, for aqueous solutions, at a temperature of 100°C. For 10 µL of solution, 20 s are suf-ficient. For larger or smaller volumes of solution, 1.5 s/µL is usually adequate time to dry the sample.

During the ashing step, volatilization of organic and low-boiling inorganic compounds occurs. The upper temperature limit for this step is the highest temperature not resulting in the loss of analyte. The major function of the ashing step is to modify the solid film on the floor of the furnace for reproducible and efficient atomization. Depending on the composition of the solid film and the specific properties of the analyte, the ashing step is carried out at temperatures between 100° and 1800°C for times ranging from 30 s to several minutes.

The atomization step is carried out rapidly, 5 to 10 s, at a temperature between 1800° and 2600°C. During this step, the thermal energy is sufficient to vaporize the ashed material and form the atomic vapor of analyte.

Electrothermal atomization has been accomplished using various refractory metals (platinum, tantalum, tungsten, etc), graphite or vitreous carbon, or metal-clad or -coated graphite fabricated into ribbons, rods, or tubes as the heating element. The graphite furnace* and the carbon rod[†] atomizers are more frequently encountered than those based on metal ribbons. These devices require inert gas sheathing to retard oxidation. In use, however, gradual oxidation leads to a loss of sensitivity and reproducibility. It is necessary, therefore, to check the calibration frequently and to replace the furnace tube or rod when these changes are observed. Depending upon the sample matrix and the atomization program, a conventional furnace tube has a typical useful life of 50 to 100 atomization cycles, and a pyrolytic coating would extend it by a factor of three.

Electrothermal atomization, like flame atomization, is vulnerable to the errors associated with matrix effects. The following "split and spike" approach is useful in identifying matrix interferences:

1. Dilute an aliquot of the sample 1 to 4 with water or appropriate solvent.

2. Spike a second aliquot with a known amount of analyte and dilute it to the same volume as the first.

3. Measure the absorbance of the two diluted aliquots and the original sample.

In the absence of matrix interferences, the concentration of

---

*Perkin-Elmer Models HGA-500 and HGA-400, Instrumentation Laboratory Model IL 655 CTF.

[†]Varian Model CRA-90, Baird Model A5100-060182.

analyte found in the original sample should agree to within ±10% of that found in the first diluted aliquot corrected for dilution, and the recovery of the spike in the second aliquot should be 90% to 110% relative to the first aliquot. If significant matrix interferences are identified, the sample should be analyzed by the method of standard additions (Section 1.5.2).

1.4. Interferences

Atomic absorption spectrometry is characterized by high sensitivity and selectivity. There are, however, several properties of the matrix, analyte, atomizer, and resonance radiation that interfere with the accurate comparison of absorbance by the sample to absorbance by the standard. These interferences are often classified as chemical, physical, and spectral, and their effects on the comparison can be minimized by matrix matching and/or background correction.

1.4.1. Chemical Interferences

Chemical interferences occur when the analyte combines with components of the matrix to form compounds that undergo atomization differently from the standards. A typical example of a chemical interference is that of phosphate from the matrix on the determination of calcium when calcium nitrate or calcium chloride standards are used. The fraction of the aspirated standard reaching the air-acetylene flame is atomized to a greater extent than that from the sample. The latter is converted to calcium pyrophosphate which is more stable than the calcium nitrate or calcium chloride at the temperature of the air-acetylene flame. Hence, when standard and sample containing the same amounts of calcium are aspirated, fewer free calcium atoms are formed from the latter and the absorbance is correspondingly lower.

This chemical interference can be minimized by employing a hotter flame or by adding reagents that form easily atomized analyte to both the sample and the standard. In the case of phosphate interference on calcium, the addition of lanthanum chloride will form lanthanum phosphate, thereby "releasing" the calcium for atomization. Similarly, the addition of disodium ethylene diamine tetraacetate (EDTA) "releases" the calcium by preventing the formation of the phosphate.

The loss of a volatile analyte compound during the drying or ashing steps in electrothermal atomization can also be classified as a chemical interference. When lead nitrate standards are used, the sample could show low results due to the loss of lead chloride formed by interaction with the matrix prior to the atomization step. Interferences of this type can be identified by splitting and spiking. The standard additions method (Section 1.5.2) can be successfully applied to the analysis of samples showing such interferences, or the matrix of such samples can be modified by chemical means to convert the analyte to the less volatile nitrate.

While not actually due to the formation of a compound, the suppression of analyte ionization by matrix elements can be considered a chemical interference. At low concentrations, a significant number of, for example, potassium atoms undergo ionization in the flame or furnace. This reduces the atomic population available for the absorption of the resonance radiation. In the presence of large amounts of sodium, the thermal ionization of potassium is suppressed. Hence, high results are obtained when serum potassium levels are measured using potassium chloride standards in pure water. To equalize the extent of ionization, an ionization buffer (in this case, a large excess of sodium chloride) is added to both samples and standards. The use of ionization buffers is recommended in the determination of alkali and alkaline earth metals, especially when high temperatures are used to overcome the phosphate interference on the latter.

1.4.2. Physical Interferences

Physical interferences are caused by differences between the physical properties affecting the nebulization/atomization of the standards and those of the samples. Both flame and furnace techniques suffer from physical interferences. Matrix matching, matrix modification, and background correction are often employed to minimize the effects of physical interferences. The standard additions method (Section 1.5.2) is also valuable in minimizing these interferences.

In conventional flame atomic absorption spectrometry, differences between the standards and samples in terms of matrix

acid, dissolved solids, or, in the case of body fluids, protein content are responsible for differences in viscosity and
surface tension. Aspiration, nebulization, and, ultimately,
atomization, therefore, take place differently for standards
and samples, and the results obtained by comparing the absorbance of the samples to those of the standards do not accurately
reflect the levels of analyte in the samples. Matrix matching
or matrix modification by solvent extraction, sample dilution,
or the addition of specific chemical reagents to both samples
and standards is a convenient remedy in many cases.

In both flame and furnace techniques, the formation of a
particular cloud or combustion product vapor from the sample
matrix during atomization gives rise to scatter or absorption
of the resonance radiation by materials other than the analyte.
This gives false absorbance. It is frequently more convenient
to compensate for false absorbance by background correction
rather than by matrix matching or matrix modification.

Background corrections are made by measuring the attenuation of both the resonance line and a nearby reference line as
they transverse the atomizer. The reference line must not be
absorbed by the analyte, and it can originate from either the
HCL or a secondary continuum source the beam of which is co-
incident with that from the HCL. The correction is made by
attributing the attenuation of the resonance line to both the
analyte and the interference and the attenuation of the reference line to only the interference. This can be done manually or electronically depending on the capabilities of the
instrumentation.

1.4.3. Spectral Interferences

Spectral interferences occur when radiation other than the
resonance line of the analyte contributes (or detracts from)
the absorbance. Modulation of the incident resonance line
from the HCL prevents emissions by the atomizer, analyte, or
matrix from affecting the absorbance. Because of the narrow-
ness of the resonance lines, there are only a few cases where
substances other than the analyte absorb within the spectral
bandwidth of the incident radiation. In such cases, the
measurements can be made at other absorption lines to eliminate this interference.

1.5. Calibration

The chief objective in making atomic absorption measurements is to relate the absorbance of the sample to its analyte
content. This is accomplished by comparing its absorbance to
that of the standard and assuming the same absorbance-concentration relationships hold for both samples and standards.
Two approaches are available for establishing the analyte content of the sample: (1) direct comparison, and (2) standard
additions.

1.5.1. Direct Comparison

The relationship developed in Section 1.2 shows a direct
proportionality between absorbance and the number of gaseous
atoms in the system. Ideally, this linear relationship should
be demonstrated when the absorbances A_1, A_2, and A_3 are plotted
against the concentrations of analyte in the corresponding
standard solutions C_1, C_2, and C_3. If proper attention has
been given to minimizing interferences, it is then possible to
use such a plot to determine the concentrations of analyte in
the samples C_x, C_y, and C_z from their absorbances A_x, A_y, and
A_z.
Absorbance-concentration plots are rarely linear over a
wide range of concentrations. Typically, such plots show
curvature toward the concentration axis at the higher concentrations. The extent of curvature and the concentration at
which it begins depend upon the amount of parasitic radiation
stimulating the detector. Even with a narrow band pass, nearby lines from the HCL transverse the atomizer. Since these
are each absorbed differently, or not at all in the case of
nonabsorbing lines, absorbance by increasing concentrations of
analyte deviate from linearity. Although current instrumentation is able to correct for curvature, it is equally feasible
to dilute the samples so that their absorbances fall in the
linear portion of the calibration curve.

1.5.2. Method of Standard Additions

Although it appears to be less convenient than direct comparison, standard additions is the method of choice when only

a few samples are to be evaluated, when the samples differ
from each other in matrix, and when the samples suffer from
undefined matrix interferences. The method of standard additions is carried out by (1) dividing the sample into several--
at least four--aliquots, (2) adding to all but the first
aliquot increasing amounts of analyte, (3) diluting all to the
same final volume, (4) measuring the absorbances, and (5)
plotting absorbance against the amount of analyte added. The
amount of analyte present in the sample is obtained by ex-
trapolating beyond the zero addition to the analyte added
axis (Figure 1.3).

In order to extrapolate, it is necessary that the additions
not exceed the linear portion of the curve. Under ideal conditions, the method of standard additions is less accurate
than direct comparison. However, it is superior to direct
comparison when severe matrix interferences are encountered.

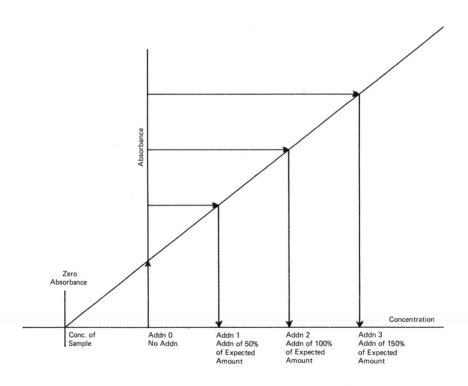

Figure 1.3. Standard addition plot

1.6. Regulatory Agencies' Requirements for Atomic Absorption Spectrometry

Many municipal, state or provincial and federal regulations require the measurement of trace element levels in environmental samples on a routine or spot basis. The Clean Water Act (CWA), the Clean Air Act (CAA), the Safe Drinking Water Act (SDWA), the Resource Conservation and Recovery Act (RCRA), and the Occupational Safety and Health Act (OSHA) all require the determination of trace elements in gaseous, liquid, and/or solid environmental and biological samples. In addition, the Food, Drug, and Cosmetic Act and the analogous state or provincial statutes may call for the determination of toxic elements in food products and consumer items. In many cases, atomic absorption spectrometry is required or recommended for these determinations.

The goal of the Clean Water Act is to achieve fishable and swimmable quality in the nation's waters. Two approaches are directed to this goal: (1) limitations on effluent discharges from industrial and municipal sources, and (2) adoption of water quality standards by the states. To achieve the former, the U.S. Environmental Protection Agency (EPA) has developed regulations for controlling discharges and requiring dischargers to apply for state and federal permits. The National (or state) Pollution Discharge Elimination System (NPDES) permits (authorized under the amendments to the Water Pollution Control Act) are used to enforce the Clean Water Act, and they require, where applicable, the determination of metals by atomic absorption spectrometry.

The goal of the Clean Air Act is to protect the public against the harmful effects of air pollution. The U.S. Environmental Protection Agency has utilized the approaches cited above to achieve these goals: (1) development of national air quality standards, and (2) adoption of state implementation plans to meet these standards. While the majority of the regulations are directed toward monitoring inorganic gasses and organic vapors, the reference method for the determination of lead in suspended particulate matter collected from ambient air specifies that the determinations of this pollutant be made by atomic absorption spectrometry.

The goal of the Occupational Safety and Health Act is to protect the worker against illness and/or injury in the occupational environment. The Occupational Safety and Health Administration has adopted numerous standards, legally enforceable regulations, governing conditions, practices, and operations to assure a safe and healthy workplace. Included among these standards are limits on airborne pollutant concentrations and methods for the determination of metals by atomic absorption spectrometry.

The goal of the Safe Drinking Water Act is to insure all citizens a safe public potable water supply. Among the many regulations established by the Environmental Protection Agency to meet this goal are maximum levels of and methodologies for the determination of inorganic and organic contaminants and specific physical properties of the water. The determination of heavy metal levels in potable water is carried out by atomic absorption spectrometry. A unique aspect of the Safe Drinking Water Act is the requirement that all determinations be carried out in certified laboratories, and the SDWA further specifies the procedures for the certification process.

The goal of the Resource Conservation and Recovery Act is twofold: (1) protection of human health and environmental quality by improving solid waste management, and (2) conservation of energy and valuable material resources. To achieve the former aspect, the regulations require that those who generate, transport, store, treat, and/or dispose of solid waste must inform the EPA of the waste's ignitability, corrosivity, reactivity, and toxicity, and, on this basis, identify it as hazardous or nonhazardous. Specific test procedures for evaluating these parameters are included in the regulations. The toxicity evaluations are made using 100 times the SDWA maximum contaminant limits, and the determinations of heavy metal contaminants are made by atomic absorption spectrometry.

The National Institute for Occupational Safety and Health (NIOSH) has developed and published procedures for the collection, dissolution, and determination of trace metals in industrial and ambient airborne material.[10] The Environmental Protection Agency has developed a procedure for the measurement of lead.[11] In both cases, the determinations are made by atomic absorption spectrometry. Similarly, the U.S. Environmental Protection Agency[9,12] and the Canadian Department of the Environment (DOE)[13] have prepared manuals for the analysis of water and waste water. These manuals contain methods for measuring trace metal levels by atomic absorption spectrometry. The U.S. Environmental Protection Agency has also published a manual for testing hazardous wastes[14] in which atomic absorption spectrometry is specified for the determination of metals. The Canadian DOE manual also specifies atomic absorption spectrometry for the determination of metals in solid environmental samples. The Canadian DOE manual and the NIOSH manual contain procedures for the determination of trace metals in biological tissues and body fluids by atomic absorption spectrometry. Federal, state, or provincial and municipal authorities frequently cite the methodologies for trace metal analysis from these manuals. Other manuals frequently cited include the American Society for Testing and Materials (ASTM) Standards[15, 16] and the American Public Health Association (APHA) Standard Methods.[17]

2. ENVIRONMENTAL SAMPLING

The reliability of any analytical result is dependent upon the integrity of the sample as well as upon the accuracy of the analytical measurement. It is necessary, therefore, that the samples taken for analysis be representative of the system under investigation, that no contamination be introduced into the samples, and that there be no changes in the compositions of the samples on standing. It must be remembered that the result of the analysis is only as good as the sampling and the sample preservation.

2.1. Site Selections for Sampling Air, Water, Solid and Liquid Wastes, and Plant and Animal Tissues

The trace element composition of biological and environmental systems varies in both time and space. It is important that these variations be considered in the planning of surveys and in the interpretation of laboratory results.

2.1.1. Site Selection for Air Monitoring

Prevailing meterological conditions affect the level of airborne material in ambient air and in the gaseous flow streams leading to the gaseous effluent outfall. Air flow rates and patterns sometimes lead to localized concentrations or depletions of the airborne material. The following considerations have been found useful in stack sampling. Every attempt should be made to sample in a section of duct showing uniform cross-sectional conditions. Such locations usually demonstrate uniformity in pressure, temperature, velocity, and composition. Locations near the existence of equipment tend to show the greatest variations in these parameters. Hence, sampling sites should be located upstream from fans, pumps, etc. Bends, enlargements, constrictions, and protrusions also lead to nonuniform conditions. Sampling sites should be located downstream from such irregularities.

2.1.2. Site Selection for Potable Water Monitoring

When sampling raw and treated water in public water supply systems, site to site variations in trace element levels are to be expected. The iron levels of samples taken at various points in the distribution system reflect both the iron levels of the source itself and that from corrosion of the water mains. Copper and zinc levels show similar site to site variations due to corrosion of the plumbing. Multiple, randomized sampling sites provide a broader overview of the system's composition than a single random sample. The number of such samples is determined by the size of the system.

2.1.3. Site Selection for Monitoring Surface Waters

When streams or rivers are sampled, careful consideration must be given to water currents and velocities. Water from the swift flow of a channel will tend to show higher trace element levels than the more placid, nearshore waters due to the higher suspended solids content of the former. During periods of heavy rainfall, however, the nearshore waters may show higher trace element levels due to surface runoff.

Stratification in lakes frequently leads to variations in many parameters for samples taken at different depths. Also, the trace element levels of nearshore waters may be significantly higher than those of the lake's center due to localized runoff or discharges.

Large bodies of water usually require the establishment of a multidimensional grid for locating sampling sites and times. Such a grid was employed to evaluate some preconstruction baseline levels of trace elements in the nearshore waters of the southern New Jersey coast.[18] Six sites were located within the square mile surrounding the proposed outfall of a multimillion gallon per day sewage treatment plant. Quarterly samples were collected 5 ft below the surface, 5 ft off the bottom, and midway inbetween at the corners of the square, at the proposed outfall in the center of the square, and at the surf line due west of the proposed outfall.

2.1.4. Site Selection for Sampling Liquid Wastes

Industrial and municipal effluents present fewer problems in selecting sampling sites than do streams and rivers. The

same considerations of flow patterns and storm water inputs
must be made, but on a much smaller scale. Usually one or two
sites per effluent stream are sufficient. These should be
located near the source of the effluent and, if a second site
is established, near the confluence of the effluent stream
and the receiving body of water. The effluent streams are
more likely to show variations in composition on a time basis
than on a site to site basis.

Sampling site selection for waste lagoons, settling tanks,
etc, follows that for lakes. The number and depth of these
sites depend upon the size of the system being sampled. Unlike the effluent streams, the waste lagoons and settling
tanks are more likely to show site to site variations.
Meterological conditions may also give rise to variations in
composition on a time basis.

The selection of sites for monitoring wells to collect
samples of landfill leacheate must consider the direction of
groundwater flow and the depth of the adjacent aquafers. In
addition to the size of the landfill, the geological, hydro-
logical, and meteorological characteristics of the site must
be carefully evaluated. The number and location of monitoring
wells is usually determined on the basis of the unique features of each and every landfill.

2.1.5. Site Selection for Sampling Solid Wastes

The identification of specific sampling sites may or may
not be necessary for solid wastes. Gross samples of such
solid wastes as dried sewage sludge or incinerator ash can be
crushed and quartered by the conventional procedures.[19, 20]
Other kinds of solid wastes require consideration on a case
by case basis.

The U.S. Environmental Protection Agency (EPA)[14] has identified six types of wastes based on the uniformity of the
processes generating them and on the degree of homogeneity
with which the contaminants are distributed within the wastes.
The six types are

1. Uniformly homogeneous

2. Nonuniformly homogeneous

3. Uniformly, randomly heterogeneous

4. Nonuniformly, randomly heterogeneous.

5. Uniformly, nonrandomly heterogeneous

6. Nonuniformly, nonrandomly heterogeneous

In general, the more homogeneous the waste, the fewer sampling sites need be identified. Table 2.1 lists several conditions in which wastes may be found and the corresponding recommended sampling points.

2.1.6. Site Selection for Sampling Plant and Animal Tissues

The requirements for selecting sites to collect plant and animal tissues depend, for the most part, on the objectives of the investigation. Obviously, a statistically significant number of specimens is required. Their trace element contents may result from their geographical locations in several ways. Localized concentrations (or depletions) of minerals or waste products can enter the food chain of both plants and animals to give some members of these species abnormally high (or low) levels of some elements. Valentine,[21] for example, has shown that some California residents have abnormally high arsenic levels in their blood, urine, and hair due to the ingestion of arsenic-containing drinking water. In assessing the levels of trace elements in plant and animal tissues it is important to recognize the possibilities of endogenous and exogenous sources. Plants grown along a heavily used highway have been shown to contain higher levels of lead than those from more remote locations.[22] Furthermore, it appears that most of this added lead burden occurs as an external deposit. Similarly, it appears that hair copper and zinc levels are due, in part, to the use of cosmetic products.[23] Gross sampling of plant and animal tissues is sufficient for evaluating environmental exposures. A more controlled sampling program is needed in work dealing with the biochemical mechanisms of mineral metabolism.

2.2. Grab Versus Composite Sampling

The mode and frequency of sampling depends upon the homogeneity of the system being sampled and upon the information

Table 2.1. Recommended Sampling Points for Wastes

Waste contaminant	Sampling points
Drum, end bung	Withdraw sample through bung opening.
Drum, side bung	Lay drum on side with bung up. Withdraw sample through bung opening.
Barrel, fiberdrum, buckets, sacks, bags	Withdraw samples through the top of the barrels, fiberdrums, buckets, and similar containers. Withdraw samples through the center of the containers and to different points diagonally opposite the point of entry.
Vacuum truck and similar containers	Withdraw sample through open hatch. Sample all other hatches.
Ponds, pits, and lagoons	Divide the surface area into an imaginary grid.[a] Take three samples, if possible: one sample near the surface, one sample at middepth or at the center, and one sample at the bottom. Repeat the sampling at each grid point over the entire site.
Waste pile	Withdraw samples through at least three different points near the top of the pile to points diagonally opposite the point of entry.
Storage tank	Sample from the top through the sampling hole.
Soil	Divide the surface area into an imaginary grid.[a] Sample each grid point

[a] The number of grid points is determined by the desired number of samples to be collected, which when combined should give a representative sample of the waste.

being sought from the system. A grab sample consists of a
single sample that reflects the characteristics of the system
at that point in time and space corresponding to the collection conditions and may be taken manually or automatically.
On the other hand, a composite sample is a mixture of several
grab samples collected from a specific sampling site over a
designated period of time. The amount of the individual sample
that is added to the total mixture should depend on the flow
of the system at the time the sample was taken. Composite
samples tend to average localized irregularities in the system. Like grab samples, composite samples can be collected
either manually or automatically. Grab samples are preferred
over composite samples when:

1. The system to be sampled is not continuous, such as
   intermittent discharges from several holding tanks.

2. The system is known to show relatively constant
   characteristics.

3. It is desired to determine whether or not the system
   shows extremes in composition.

In general, the analysis of a large number of samples collected
at different times from several different sites in the system
provides much more information than could possibly be obtained
from the analysis of a single composite sample. (The victims
of Minamata disease find little consolation in learning that
the average mercury concentration in fish is below 0.5 ppm.)
The single major advantage of composite sampling is that it
generally reduces the laboratory work load, with corresponding
cost reductions.

2.2.1. Air Samples

The collection of samples in airborne materials for elemental analysis frequently takes place over time intervals
ranging from 2 h[24] to 2 weeks.[25] The development of high-
volume air samplers was responsible for the reduction to the
lesser value. Even though the samples are collected over prolonged periods of time, they are usually considered as grab
samples on a site by site basis.[26]

2.2.2. Water and Wastewater Samples

The elemental composition of water and waste water is best
evaluated from grab samples taken from multiple sites on a
routine basis.[12, 15, 27] The number of sampling sites is determined by the size of the system, and the frequency of sampling depends upon the variability of its composition.

2.2.3. Solid Samples

Solid waste samples, depending on their particular origin
and on the information sought, are most conveniently collected
as composites. Homogeneity of such samples is essential.
When homogeneity is lacking, grab sampling becomes necessary.

2.2.4. Plant and Animal Tissue Samples

Both grab sampling and composite sampling have been used to
collect plant and animal tissues. Grain samples for the determination of arsenic, seafood samples for the determination
of mercury,[28] and milk-free infant formula for the determination of chromium[29] were composited prior to analysis. Grab
sampling is used to collect blood for the determination of
lead. In general, composite sampling is used to collect materials for elemental analysis in food processing or other pro-
duction line operations. Grab sampling is used in collecting
materials of clinical or laboratory significance.

2.3. Sample Containers

The sample container functions to hold the sample from
the time of collection to the time of analysis. During this
period, the container may neither add to nor take away from
the elemental composition of the sample, and the container
must protect the sample from contamination and/or loss.

2.3.1. Containers for Air Samples

With few exceptions, samples of airborne materials are collected by drawing measured volumes of air through nitrocellu-
lose-cellulose acetate filters or through glass fiber filters.

Table 2.2. Typical Purity of Cellulose Ester (CE)
and Glass Fiber (GF) Filters

Element	CE (ppm)[a]	GF (ppm)[a]	Element	CE (ppm)[a]	GF (ppm)[a]
Al	2	X	Mg	1.8	X
Ag	ND	<0.02	Mn	<0.05	1.2
B	<2	X	Mo	ND	0.5
Ba	<1	X	Na	40	X
Be	ND	0.3	Ni	ND	2.5
Bi	ND	1.0	Pb	0.2	2.5
Ca	13	X	S	<5	ND
Cd	ND	0.5	Sb	<0.02	ND
Co	ND	0.3	Si	<2	X
Cr	0.3	1.2	Sn	ND	2.5
Cu	0.4	0.7	Ti	ND	2.5
Fe	6	25	V	ND	2.5
K	1.5	X	Zn	0.6	25

[a] ND, not detected; X, major constituent of filter.

The loaded filters are sealed in their original cassettes during the period from collection to analysis. Typical background or blank values for commercially available cellulose ester and glass fiber filters are presented in Table 2.2.

2.3.2. Containers for Liquid Samples

The regulatory and advisory agencies have defined suitable containers for most water and wastewater samples. The New Jersey Department of Environmental Protection (NJDEP)[27] specifies and the U.S. EPA,[8] the American Society for Testing and Materials (ASTM),[15] and the American Public Health Association (APHA)[17] recommend chemically resistant glass or chemically resistant polyethylene bottles and closures with inert liners. Moody and Lindstrom[30] have evaluated several different plastic containers for trace element samples, and

Laxen and Harrison[31] have compared several different methods for cleaning polyethylene containers for freshwater samples.

2.3.2.1. Cleaning of Liquid Sample Containers   The following procedure[32] has been recommended for cleaning sample containers prior to use:

1. Wash the bottles and closures with a good quality detergent in hot water.

2. Rinse the bottles and closures well with tap water.

3. Rinse the bottles and closures with 1:1 nitric acid.

4. Rinse the bottles and closures with tap water.

5. Rinse the bottles and closures with successive portions (10-15) of high-purity water.

6. Invert the bottles and closures in a dust-free environment for drying.

Bottles prepared in this manner show negligible blank values when filled with high-purity water and evaluated at a later date. Data for trace elements leached from borosilicate glass by high-purity water after 2 weeks at 50°C were as follows: B, 67 µg/L; Si, 57 µg/L; Al, 27 µg/L; Cu, 27 µg/L; Mg, 8 µg/L; and Na, 630 µg/L.[33]

For water and wastewater samples, the U.S. Environmental Protection Agency[9] requires, "All glassware, linear polyethylene, polypropylene or Teflon containers, including sample bottles, should be washed with detergent, rinsed with tap water, 1:1 nitric acid, tap water, 1:1 hydrochloric acid, tap water, and deionized water, distilled water, in that order. This procedure is also acceptable for the preparation of containers for hazardous waste samples.[14]

2.3.3. Containers for Solid Samples

Solid waste samples are conveniently collected in wide mouth plastic or glass jars cleaned as outlined above. One-liter polyethylene wide mouth jars with linerless polyethylene caps were used to collect samples of sediments and soils.[34] The jars were washed with detergent, thoroughly rinsed, and

washed in 1:1 nitric acid and then in 1:1 hot hydrochloric
acid. Each wash was followed by thorough rinsing with deionized, distilled water. Jars serving as shipping blanks were
sent into the field, returned empty, and rinsed with nitric
acid. Analysis of the nitric acid rinses showed nondetectable
levels of aluminum, silver, barium, beryllium, bismuth, cadmium, cobalt, copper, molybdenum, nickel, lead, antimony,
selenium, tin, strontium, titanium, and vanadium. The following elements were found in microgram amounts: calcium,
117; chromium, 16; iron, 108, potassium, 10; magnesium, 7;
manganese, 3; sodium, 36; and zinc, 15. If the jars had been
filled to capacity, and if all of the acid-leached elements
contaminated the sample, the effect of the added calcium
would have been less than 0.1 ppm, and the effect of the added
manganese would have been less than 1 ppb.

2.3.4. Containers for Plant and Animal Tissues

Wide mouth plastic and glass containers have been used for
plant and animal tissues. Blood samples are routinely collected in evacuated tubes. Nackowski,[35] et al have made a
survey of the trace metal contaminants in the blood collection
tubes of several manufacturers. Their study indicates that
zinc, lead, and cadmium contamination of blood samples can be
a significant problem with certain blood collection tubes
under normal handling, shipping, and storage conditions.
Lecomte et al[36] have found significant zinc contamination of
blood samples collected in one particular brand of evacuated
tubes. Handy[37] identified the rubber stoppers of these tubes
as the major source of zinc contamination, and Gorsky and
Dietz[38] recommend keeping the blood sample in a syringe to
prevent contamination.

2.4. Sample Collection

2.4.1. Air Sampling

Adequate sampling of the atmosphere is often difficult because the particles suspended in the atmosphere are not uniformly dispersed. They do not follow the gas stream lines in
the atmosphere, and they are frequently subjected to forces
other than aerodynamic ones. Before any analysis of samples

is performed for purposes of environmental monitoring, careful sampling must be conducted.

Adequate sampling of particulate matter involves, in most cases, isokinetic sampling (particles enter the sampling orifice without undergoing a change in velocity). Anisokinetic sampling leads to the collection of a disproportionately small number of the larger particles. Such sampling errors are minimal in the open atmosphere. When samples are collected from stacks or ducts, the errors associated with anisokinetics are of serious concern. The concentration of some trace elements is enriched in or on the smaller particles.[39] Analysis of a sample collected anisokinetically from a stack would show erroneously high values for these elements in the particulate emissions.

The sampling of airborne particulates is most frequently carried out by filtration. Inertial collection devices add the dimension of selectivity in retaining particles of specific size ranges. For subsequent chemical analyses, readily soluble filter media are frequently selected. Sampling from high-temperature environments requires the use of alundum or fiber glass filters. Some typical filter recommendations are tabulated in Table 2.3. The single-stage impactor (wet impinger) is suitable for collecting airborne particulates from a variety of environments, and the multistage impactor (cascade impactor) retains the larger particles in the earlier stages, thereby permitting simultaneous collection and size sorting.

2.4.2. Potable Water Sampling

Sampling potable water supplies may involve the collection of water samples that are representative of the raw non-treated water (raw water samples), treated finished water (plant delivered samples), and distribution system (system samples). When collecting system samples, the water should be allowed to run from a spigot to waste until a volume equal to that in the service line has flushed through the collection point. Water softeners, aerators, etc, must be removed prior to flushing and sampling. Similarly, plant delivered samples are collected immediately following treatment, and raw water samples are collected prior to treatment. Glass or plastic containers prepared as described previously are filled from the appropriate locations.

Table 2.3. Filter Recommendations for Air Sampling

Element	Filter[a]	NIOSH reference[b]
Antimony compounds	MCE 0.8 μm	S2, 261, 189
Arsenic	MCE 0.8 μm	139, S309, 188, 180
Barium compounds	MCE 0.8 μm	S198
Beryllium	MCE 0.45 μm	121, S339
Boron oxide	PVC 2 μm	S349
Cadmium dust	MCE 0.8 μm	S312
Cadmium fumes	MCE 0.8 μm	S313
Calcium oxide	MCE 0.8 μm	S205
Chromates	PVC 5 μm	169, S317
Chromium compounds	MCE 0.8 μm	S352, S323
Cobalt compounds	MCE 0.8 μm	S203
Copper compounds	MCE 0.8 μm	S186
Copper fumes	MCE 0.8 μm	S354
Platinum	MCE 0.8 μm	S191
Rhodium	MCE 0.8 μm	S188
Selenium compounds	MCE 0.8 μm	S190, 181
Silver compounds	MCE 0.8 μm	173
Tantalum	MCE 0.8 μm	—
Tellurium	MCE 0.8 μm	S204
Thallium compounds	MCE 0.8 μm	S306
Tin compounds	MCE 0.8 μm	S183, 176
Titanium compounds	MCE 0.8 μm	S385
Tungsten compounds	MCE 0.8 μm	271
Uranium compounds	MCE 0.8 μm	—
Vanadium dust	MCE 0.8 μm	S391
Vanadium fumes	MCE 0.8 μm	S388
Vttrium	MCE 0.8 μm	S200
Zinc chloride fume	MCE 0.8 μm	22 SDS
Zinc oxide fume	MCE 0.8 μm	173
Zirconium compounds	MCE 0.8 μm	S185, 250

[a] MCE, mixed cellulose esters; PVC, polyvinyl chloride.

[b] NIOSH Manual of Analytical Methods, Vol. 1 - Vol. 6, U.S. Government Printing Office (GPO).

2.4.3. Groundwater Sampling

When collecting samples from municipal wells, industrial wells, domestic wells, or monitoring wells, flushing or pumping of the well is almost always required to insure a representative sample of the ground water. Before collecting the sample, pump to waste a volume of water equal to three times that standing in the casing of the well. The number of gallons of water per foot of casing of diameter d is equal to:

$$(d/2)^2/19.25$$

Devices needed for pumping or collecting water from wells may include: portable centrifugal pumps, 3-in. or 4-in. submersible pumps, bailers, or Kemmerer-type samplers. Once the well has been flushed, samples are then collected in the appropriate containers.

2.4.4. Wastewater Sampling

Municipal and industrial waste waters may be collected as either grab or composite samples depending on the purpose of the monitoring. When the monitoring is required for federal or state compliance, the type of sample and location of collection is often specified by the appropriate agency. This may be indicated on the National Pollutant Discharge Elimination System (NPDES) permit which is issued by the EPA. Operating personnel familiar with the plant operation should be available to identify influent and effluent streams, equalization tanks, etc. Samples should be taken below the surface of the stream, lagoon, pond, etc, to avoid surface contamination.

Whenever possible, the sample should be collected directly into the sample container. This may be accomplished by using such devices as the Wheaton grab sampler. This sampler will conveniently handle 1-L bottles at distances of up to 6 ft.

When monitoring municipal and industrial discharges, composite sampling is the method of sample collection that is often employed. This provides a sample that is representative of the discharge over a defined period of time (ie, 24 h). Composite samples may be collected as hourly, manually collected, grab samples over a 24-h period or an automated composite sampler may be used at a cost savings in long-term monitoring programs (Figure 2.1).

Figure 2.1. Composite sampler. Reproduced with permission
from ISCO, Inc. Courtesy Mr. Waddell.

2.4.5. Surface Water Sampling

Stream and river sampling methods should be dictated by
the information being sought from the monitoring survey. Grab
samples using the Wheaton grab sampler,[40] a Kemmerer sampler,
or even by directly immersing the sample bottle into the stream
provide an inexpensive means of collecting water samples for
spot checking streams for trace elements (Figures 2.2 and 2.3).

Figure 2.2. Stream sampling.

Figure 2.3. Stream sampling equipment.

When it is necessary to collect a representative sample of a
stream or river for the analysis of trace elements in both
dissolved and suspended material, the U.S. standard sampler*
is used to collect a composite sample representing a cross
section of the water body at a single point in time. This
sampling device is often used when determining sediment transport and stream wasteload allocations.

   The U.S. standard sampler allows water and suspended material to enter the sample bottle at the same velocity as the
surrounding stream velocity. In effect, the particulate matter is sampled isokinetically, and the sample collected
closely reflects the actual ratio of solid to liquid in the
system under investigation. To collect a composite sample

---

*U.S. standard samplers were developed by the Federal Inter-
 Agency Sedimentation Project (FIASP) of the Inter-Agency
 Committee on Water Resources located at St. Anthony Falls
 Hydraulic Laboratory in Minneapolis, Minnesota.

representative of a stream, a transect is first made of the stream (using a measuring tape) and several equidistant points in the transect are selected. At each point, a verticle composite sample of the water column is collected using the U.S. standard sampler and all of the verticle composites are combined to make one composite sample that is representative of the cross section of the stream. Wadeable streams may be sampled using hand-held samplers, while nonwadeable streams and rivers may be sampled from a boat or bridge using a crane operated sampler.[41]

Streambed samples are collected upstream of the individual doing the collection to avoid interfering with the water flow and to prevent disturbing the stream bed. Such samples are usually taken from the upper 1 or 2 in. of the sediment lying on the stream bottom.

Lake, estuarine, and/or marine water samples are usually collected as hand-held grab samples or when samples at specific depths are needed, a Kemmerer sampler may be used. When using a boat in the collection of these samples, the sampling site should be approached from the down current direction and the sample should be collected from the front of the boat.

2.4.6. Solid Waste Sampling

Specific procedures for sampling solid wastes are described in the Federal Register.[42] Kratochvil and Taylor[43] have presented an excellent overview of the fundamental aspects of sampling solid materials, and the Bendix Corporation[44] has developed an audiovisual training program entitled "Sampling for Toxic Substances."

The sampling equipment recommended for various types of samples found under a variety of environmental conditions is listed in Table 2.4, and the number of samples to be collected under these conditions is summarized in Table 2.5.

2.4.7. Sampling Plant and Animal Tissues

Samples of plant and animal tissue collected under natural conditions are always contaminated with the medium in which they live. Particles of soil adhere to the roots of land plants, and sections of animal tissue are filled with blood

and/or other fluids. While it may be possible to wash plant
and animal tissues free of contamination from their natural
media, such procedures are also likely to leach trace elements
out of the tissues. Red blood cells cannot be washed in high-
purity water. They undergo hemolysis. Such tissues can be
successfully washed in isotonic solutions made from high-
purity salts and high-purity water. The samples, of course
become heavily contaminated with the salts used to maintain
the osmotic pressure.

Tissue samples are vulnerable to contamination during collection. Talc from the pathologists' gloves is frequently a
source of trace element contamination. Wilkerson et al[45] have
shown that cobalt, chromium, and iron can be leached from talc
by body fluids. Particles of talc falling into the sample,
however, present a much more serious contamination problem.
The implements used to collect the samples are also potential
sources of contamination. Versieck and Speecke[46] have shown
that blood is contaminated with manganese, copper, chromium,
iron, nickel, cobalt, and zinc from disposable venipuncture
needles during collection. They similarly demonstrated that
liver biopsies were contaminated with manganese, copper,
scandium, chromium, copper, nickel, cobalt, zinc, silver,
antimony, and gold from the disposible surgical blades used
to collect the samples. The extent of manganese contamination
of the blood serum was at a level equal to its normal manganese content. Plastic instruments should be used in handling
the specimens to eliminate contamination by trace metals.
Knives can be fabricated from high-purity quartz or from a
metal other than those for which the analyses will be carried
out. High-purity titanium has been found useful for this
purpose.

The International Atomic Energy Agency (IAEA) has recommended the following procedure for the collection of biological specimens:[47]

> Since the elements of interest are present in the tis-
> sues at concentrations of a few micrograms per gram
> and even much lower, great care is needed to avoid
> metal contamination. Handling of the samples should
> therefore be kept to a minimum and metal free plas-
> tic gloves should be worn. No chemical fixatives
> may be used and the sample should not be pierced
> through with a metal instrument, nor should they be

Table 2.4. Sampling Equipment for Various Samples

Waste types	Sampling points								
	Drum	Sacks, bags	Open-bed truck	Closed-bed truck	Storage tanks or bins	Waste piles	Ponds, lagoons, and pits	Conveyor belts	Pipe
Free-flowing liquids	Coliwasa			Coliwasa	Weighted bottles		Dipper		Dipper
Sludges	Trier		Trier	Trier	Trier				
Moist powders	Trier	Trier	Trier	Trier	Trier	Trier	Trier	Shovel	
Dry powders	Thief	Thief	Thief	Thief	Thief	Thief	Thief	Shovel	
Sand or packed powders	Auger	Auger	Auger	Auger					
Large-grained solids	Large trier	Large trier	Large trier	Large trier	Large trier	Large trier	Large trier		

Table 2.5. Number of Samples To Be Collected

Case no.	Information desired	Waste type	Container type	Number of samples to be collected
1	Average concentration	Liquid	Drum, vacuum truck, or similar	1 collected with coliwasa
2	Average concentration	Liquid	Pond, pit, or lagoon	1 composite sample of several samples collected at different sampling points or levels
3	Average concentration	Solid powder	Bag, drum, bin, or sack	Same as case no. 2
4	Average concentration	Waste pile		Same as case no. 2
5	Average concentration	Soil		1 composite sample of several samples collected at different sampling areas
6	Concentration range	Liquid	Drum, vacuum, truck, storage tank	3 to 10 separate samples each from a different depth of the liquid

7	Concentration range	Liquid	pond, pit, or lagoon	3 to 20 separate samples from different sampling points and depths
8	Concentration range	Solid powder	Bag, drum, bin	3 to 5 samples from different sampling points
9	Concentration range	Waste pile		Same as case no. 8
10	Concentration range	Soil		3 to 20 separate samples from different sampling areas
11	Average concentration for legal evidence	All types	All	3 identical samples or 1 composite sample divided into 3 identical samples if homogeneous
12	Average concentration	Liquid	Storage tank	Same as case no. 2
13	Average concentration	Liquid	Storage tank	Same as case no. 6

rinsed with tap water or any other medium. The instruments used for handling the samples prior to the
analysis, or for cutting or breaking them into small
pieces, may be a potent source of contamination unless the proper precautions are taken. Therefore,
an autopsy collection kit comprising plastic forceps, precleaned plastic vials, specially prepared
titanium knives and a stone for sharpening them, are
prepared at the IAEA and supplied to the collabor-
ating pathologists. Silica or plastic knives have
also been tested and found to introduce no significant contamination at least for certain elements
and therefore their use is acceptable. If it is
necessary to use steel instruments those should
preferably be of carbon steel rather than stainless
steel. Stainless steel instruments, for example
forceps, scalpels and scissors, could be used, but
definitely not when chromium is to be analyzed.
The danger of contaminating samples with chromium
from stainless steel instruments and of course even
more from chromium-plated instruments is well documented. As compared with past experience with glass
knives, quartz knives or polyethylene knives such as
those used in picnic sets or stiffened by cooling in
liquid nitrogen, the titanium knives recently produced at IAEA are well acceptable to pathologists,
have a low degree of metal contamination, can be
sharpened and are easy to produce.

   Although blood is not one of the tissues of primary interest in the present program, some consideration has been given to the problem of obtaining
suitably uncontaminated samples. Blood collecting
needles made from nickel or platinum on Teflon®
mounts are commercially available and are considered to be suitable for this purpose. For uniformity, the IAEA recommends the following be observed
in collecting samples:

Tissue	Sample
Heart	Anterior wall of the left ventricle.
Liver	Superior, anterior surface of right lobe.

Tissue	Sample
Kidney	Cortex of lower pole of left kidney.
Hair	Lock from occipital region
Toenails	Wide clipping from right and left great toes.

2.5. Sample Preservation and Holding Times

It is necessary that the sample undergo no changes in composition during the interval between collection and analysis. In some instances, partial loss of analyte is prevented by the addition of preservatives. In others, the samples are frozen or cooled to retard such losses. Typical of such losses are adsorption of metal ions from water samples on to the walls of glass containers and the spontaneous precipitation of calcium phosphate with coprecipitation of other metals from urine samples on standing. Acidification, cooling, and prompt analysis generally minimize such losses.

2.5.1. Preservation of Air Samples

Airborne particulates filtered from air samples are usually quite stable with respect to trace element contents. Thompson,[48] however, has indicated that fly ash can absorb mercury from laboratory air containing mercury vapor. The National Air Surveillance Network has some 200,000 samples, some collected over 20 years ago, stored in its sample bank repository. Air samples collected in impingers are also stable. Analysis, nevertheless, should not be delayed for prolonged periods of time.

2.5.2. Preservation of Water Samples

Both the NJDEP and the U.S. EPA require the water and wastewater samples for trace metal analysis be preserved by the addition of concentrated nitric acid until the pH of the resulting solution is reduced to at least 2 immediately following sample collection. Samples so preserved may be held up to a maximum of 6 months prior to analysis for aluminum, antimony, arsenic, barium, beryllium, cadmium, total chromium, copper, iron, lead, manganese, nickel, selenium, silver, tin, zinc, calcium, magnesium, potassium, and sodium. The determination of hexavalent chromium must be initiated within 24 h of collection, and mercury analysis may not be delayed more than 38 days when the preserved samples are stored in glass bottles or more than 13 days when plastic bottles are used.

Subramanian et al[49] have studied the loss on storage of 11 trace metals from both natural and synthetic water samples. Pyrex glass, Nalgene polyethylene, and Teflon containers were used. Trace metal loss from solution was evaluated as a function of time in the pH range of from 1.5 to 8.0. Acidification to below pH 1.5 with nitric acid and storage in the Nalgene containers appeared to be the most efficient in preventing loss of trace metals from natural water.

Truitt and Weber[50] found that significant amounts of copper were lost when water at pH 7 was passed through cellulose acetate filters on glass supports. Losses were minimal when acid washed, polycarbonate filters on polycarbonate supports were used. Hoyle and Atkinson[51] have reported that diammonium EDTA prevented the absorption of lead and cadmium from aqueous solutions of pH 10 on to the glass or plastic surfaces for as long as 40 days. Das et al[52] have recently presented the quantitative aspects of adsorption of metal ions from aqueous solutions to container surfaces.

2.5.3. Preservation of Solid Waste Samples

The preservation of solid wastes is ill defined. Refrigeration is frequently recommended. Given the wide array of possible material, the only generalization possible is a minimum of delay in preparing the samples for analysis.

2.5.4. Preservation of Plant and Animal Tissue Samples

The traditional preservatives for plant and animal tissues are to be avoided. Formalin and heparin are significant sources of contamination. Bowen[41] has reported that the trace metal impurities in commercial heparin are frequently higher than the normal levels of these elements in human blood. These values are compared in Table 2.6.

The IAEA recommends that tissue samples be frozen immediately

Table 2.6. Trace Elements in Heparin and Human Blood

Element	ppm in heparin	ppm in blood
Barium	2.5-12	0.069
Calcium	300-2900	62
Copper	0.65	1.1
Manganese	3.6	0.026
Strontium	5-92	0.039
Zinc	28	6.5

after collection, and that they be kept frozen until their preparation for analysis. Freeze drying and vacuum drying have been found useful in preserving tissue samples. The samples, nevertheless, should be prepared for analysis without delay.

3. SAMPLE PREPARATION

Samples for elemental analysis by atomic absorption spectrometry are usually introduced to the spectrometer as
liquids. Hence, airborne particulates, solid wastes, and
plant and animal tissues must be dissolved by suitable means.
Because atomic absorption spectrometry is a comparative technique, the samples and the standards require matrix matching.
Consequently, both liquid and solid samples are frequently
treated by chemical means to define the matrix. While atomic
absorption spectrometry is sensitive and selective, additional
chemical procedures are sometimes necessary to concentrate
and/or isolate the analyte. In the course of these procedures,
as well as those used to dissolve the sample or change its
matrix, extreme caution must be exercised to prevent contamination. Equal care must be taken to insure against the loss
of analyte.

3.1. Decomposition and Solubilization Methods

Decomposition involves removing the organic matrix by con-
verting it to compounds that are easily volatilized. The inorganic residue is then treated with appropriate solvents to
complete solution. Among the more common decomposition methods are digestions with acids and combustion with oxygen or
appropriate fluxes. The former are vulnerable to contamination by impurities in the acids, and the latter are subject
to losses by volatilization as well as contamination when
fluxes are used.

3.1.1. Acid Digestion (Wet Oxidation) Procedures

Sample preparation frequently begins with acid digestion.
Filters containing airborne particulates, soils, sediments,
sludges, liquids containing suspended solids, and plant and
animal tissues are dissolved by decomposing the sample with
acid under a wide variety of experimental conditions.

3.1.1.1. Air Samples   Practically all of the procedures
for preparing airborne particulate material begin with the
acid digestion of the filter that served to collect the sample.
Cellulose ester filters are usually dissolved in nitric acid.
Sometimes, hydrochloric acid, sulfuric acid, and/or perchloric
acid is added to aid in the destruction of organic material.
Hydrogen peroxide has also been used for this purpose. Hydrofluoric acid is required to dissolve glass fiber filters.

The NIOSH Manual of Analytical Methods[10] recommends nitric
acid for decomposing membrane filters and dissolving the metals
present in the sample of airborne particulates. The procedure
for the preparation of such samples is as follows.

> The filters containing the samples of airborne particu-
> lates and filter blanks are transferred to clean
> 125-ml Phillips or Griffin breakers and 6.0 ml of
> nitric acid is added. Antimony samples are ashed
> in 6.0 ml of a 5:1 mixture of nitric acid and sul-
> furic acid. Each beaker is covered with a watch
> glass and heated on a hot plate, 140°C, in a fume
> hood until the sample dissolves and a slightly yel-
> low solution is produced. Approximately 4 hours of
> heating will be sufficient for most air samples.
> However, subsequent additions of nitric acid may be
> needed to completely ash and destroy high concentra-
> tions of organic material, and under these conditions
> longer ashing times will be needed. Once the ashing
> is complete, as indicated by a clear solution in the
> beaker, the watch glass is removed and the sample is
> allowed to evaporate to near dryness, approximately
> 0.5 ml.
>
> Remove the beaker from the hot plate, cool, and
> add 1 ml of nitric acid and 2 to 3 ml of distilled
> water. For lead samples, concentrated hydrochloric
> acid is used instead of nitric acid. The solution
> is quantitatively transferred with distilled water
> to a 10-ml volumetric flask. If any elements being
> determined require the ionization buffer, 0.2 ml of
> 50 mg/ml Cs is added to the volumetric flask. If
> any elements requiring the releasing agent are being
> determined, 0.2 ml of 50 mg/ml La is added to each
> volumetric flask. The samples are then diluted to
> volume with water.

The 10-ml solution may be analyzed directly for any element of very low concentration in the sample. Aliquots of this solution may then be diluted to an appropriate volume for the other elements of interest present at higher concentrations. The dilution factor will depend upon the concentration of elements in the sample and the number of elements being determined by this procedure.

The EPA "hot extraction procedure"[11] makes use of nitric acid to dissolve trace elements, particularly lead, from airborne particulates collected on large glass fiber filters. The following is a condensation of this procedure.

Cut a 3/4 inch by 8 inch strip from the exposed filter. Fold the strip in half twice and place it in a 150-ml beaker. Add 15 ml of 3 M nitric acid to cover the sample. Cover the beaker with a watch glass.

Place the beaker on the hot plate contained in a fume hood, and boil gently for 30 minutes. Do not let the sample evaporate to dryness.

Remove the beaker from the hot plate and cool to near room temperature. Quantitatively transfer the sample as follows:

Rinse the watch glass and sides of the beaker with deionized water.

Decant the extract and rinsings into a 100-ml volumetric flask.

Add deionized water to the 40-ml mark on the beaker.

Cover with the watch glass, and set aside for a minimum of 30 minutes. This critical step cannot be omitted: It allows the nitric acid trapped in the glass fiber filter to diffuse into the rinse water.

Decant the water from the filter into the volumetric flask.

Rinse the filter and the beaker twice with deionized water and add the rinsings to the volumetric flask.

Stopper the flask and shake it vigorously. Set aside for approximately 5 minutes or until foam has subsided.

Bring the solution to volume with deionized water.

Mix thoroughly.

Set aside for one hour prior to analysis to allow suspended matter to settle.

If sample is to be stored for subsequent analysis, transfer to a linear polyethylene bottle.

Zdrojewski and co-workers[53, 54] have dissolved glass fiber filters in hydrofluoric acid-nitric acid mixtures to prepare the sample for the determination of lead in airborne particulates. Cellulose ester filters with samples for lead analysis have been dissolved in nitric acid by Janssens and Dams,[55] in nitric acid-hydrochloric acid mixture by Severs and Chambers,[56] in nitric acid-hydrochloric acid-perchloric acid mixture by Gallorini et al,[57] and in nitric acid with hydrogen peroxide by Szivos et al.[58]

3.1.1.2. Water Samples and Wastewater Samples   Samples of water and waste water are frequently subjected to acid digestion to dissolve suspended material, to destroy dissolved organic material, and to define the anionic medium for matrix matching.

The EPA Manual of Methods for Chemical Analysis of Water Wastes[8, 9] describes a nitric acid digestion procedure to prepare samples for the determination of total metals. The procedure is the following:

For the determination of total metals the sample is acidified with 1:1 redistilled nitric acid to a pH of 2 at the time of collection. The sample is not filtered before processing. Choose a volume of sample appropriate for the expected level of metals. If much suspended material is present, as little as 50-100 ml of well mixed sample will most probably be sufficient.

Transfer a representative aliquot of the well mixed sample to a Griffin beaker and add 3 ml of concentrated redistilled nitric acid. Place the beaker on a hot plate and evaporate to dryness cautiously, making certain that the sample does not boil. Cool the beaker and add another 3 ml portion

of concentrated, redistilled nitric acid. Cover the
beaker with a watch glass and return it to the hot
plate. Increase the temperature of the hot plate so
that a gentle reflux action occurs. Continue heating, adding additional acid as necessary, until the
digestion is complete (generally indicated by a
light colored residue). Add sufficient distilled
1:1 hydrochloric acid and again warm the beaker to
dissolve the residue. Wash down the beaker walls
and watch glass with distilled water and filter the
sample to remove silicates and other insoluble material that could clog the atomizer. Adjust the volume to some predetermined value based on the expected
metal concentration. The sample is now ready for
analysis. Concentrations so determined shall be reported as "total." STORET parameter numbers for
reporting this type of data have been assigned and
are given for each metal.

  Certain metals such as titanium, silver, mercury,
and arsenic require modification of the digestion
procedures and the individual sheets for these
metals should be consulted.

The 1980 Annual Book of ASTM Standards, Part 31, Water,[15]
calls for a more mild digestion procedure in which a 100-mL
water sample is treated with 5 mL of hydrochloric acid and
heated below the boiling point until the volume of the solution is reduced to 20 mL prior to the determination of most
metals. For the determination of antimony and arsenic, sulfuric acid-nitric acid mixture is used instead of hydrochloric
acid.

Hydrofluoric acid has been used to digest water samples
taken from the Great Lakes prior to the determination of trace
element levels by flameless atomic absorption spectrometry.[59]

3.1.1.3. Liquid Waste Samples   The EPA has recommended a
nitric acid-hydrogen peroxide digestion procedure to prepare
industrial effluent samples for trace metal analysis by flameless atomic absorption spectrometry.[60]   A similar procedure
has been recommended by the EPA for the decomposition of

sewage sludge prior to elemental analysis.[61] The procedure
is as follows:

> Weigh and transfer to a 125 ml conical Phillips' beaker
> a 1.0 gram portion of the sample which has been
> dried at 60°C, pulverized, and thoroughly mixed.
> Add 5 ml of 1:1 nitric acid and cover with a watch
> glass. Heat the sample at 95°C and reflux to near
> dryness. Allow the sample to cool, and add 4 ml of
> concentrated nitric acid and again reflux to near
> dryness. After the second reflux step has been com-
> pleted and the sample has cooled, add 1 ml of 1:1
> nitric acid and 3 ml of 30% hydrogen peroxide. Re-
> turn the beaker to the hot plate for warming to
> start the peroxide reaction. Care must be taken
> with the start of effervescence that losses do not
> occur or the reaction is not too vigorous. Heat
> until effervescence subsides and cool the beaker.
> Continue the addition of 30% hydrogen peroxide in
> 1 ml aliquots with warming until the effervescence
> is minimal or the general sample is unchanged.
>
> If the sample is being prepared for the furnace
> analysis of Sb and/or direct aspiration analysis of
> Sb, Be, Cd, Cr, Cu, Pb, Ni, and Zn, add 2 ml of 1:1
> hydrochloric acid, return the covered beaker to the
> hot plate and reflux for an additional 10 minutes.
> After cooling, filter through Whatman No. 42 filter
> paper (or equivalent) and dilute to 50 ml with de-
> ionized water. The sample is now ready for analysis.
>
> If the sample is being prepared for the furnace
> analysis of As, Be, Cd, Cr, Cu, Pb, Ni, Se, Ag, Tl
> and Zn, or the direct aspiration analysis of Ag and
> Tl, add 1 ml of 1:1 nitric acid, return the covered
> beaker to the hot plate and reflux for an additional
> 10 minutes. After cooling, filter through Whatman
> No. 42 filter paper (or equivalent) and dilute to
> 50 ml with deionized distilled water. To prepare
> the analysis solution, withdraw an aliquot, add any
> required reagent or matrix modifier and dilute with
> deionized distilled water to twice the volume of the
> aliquot withdrawn. The sample is now diluted in 1%
> nitric acid and ready for analysis.

The EPA has also offered the following sulfuric acid-nitric acid digestion to prepare oil-, grease-, or wax-containing wastes for elemental analysis by atomic absorption spectrometry.[14]

> Weigh out a 100 gm representative sample of the waste or extract. Separate the phases, if more than one is present, and weigh each phase.
>
> Weigh 2.0 gms of the organic phase into the digestion or Kjeldahl flask. Add 10 ml of sulfuric acid and a 6 mm glass bead. Swirl flask to mix the contents.
>
> If using a Kjeldahl flask, approximately 3/4 of the neck of the flask should be cooled by air by directing an air stream against the neck of the flask. If using the flask and condenser apparatus, connect the Allihin condenser and circulate cooling water.
>
> Heat the flask gently and continue heating until dense white fumes appear. While boiling, cautiously add 1 ml of nitric acid dropwise to oxidize the organic material. This may be done through the condenser. When the nitric acid has boiled off and dense white fumes reappear repeat the treatment with an additional 1 ml of nitric acid. Continue the addition of nitric acid in 1 ml increments until the digestion mixture is no darker than a straw color, indicating that the oxidation of the organic material is almost complete.
>
> Cool the flask slightly and add 0.5 ml (dropwise) of hydrogen peroxide. Heat until dense white fumes appear, and while boiling cautiously add 1 ml of nitric acid dropwise. When the nitric acid has boiled off and dense white fumes reappear repeat the treatment with hydrogen peroxide and nitric acid until the digestion mixture is colorless, at which time the organic material will be completely oxidized. Four treatments will usually suffice. The total amount of peroxide should be noted.
>
> When oxidation is complete, allow the flask to cool, wash down the mouth, neck/condenser with a small volume of distilled water (5 ml) and mix the

contents. Continue heating to the appearance of
dense white fumes.

Cool and dilute to a total volume of 25 ml. Proceed with the determination of metals.

3.1.1.4. Samples of Soils, Sediments, and Sludges   Soils,
sediments, and sludges have been subjected to a wide variety
of acid digestion procedures. Khalily[62] and Wisseman and
Cook[63] have digested sediment samples with nitric acid to
solubilize trace metals. Theis et al[64] have used aqua regia
and hydrofluoric acid to dissolve soil samples; and Agemian
and Chau[65] have employed this mixture to prepare sediments for
atomic absorption spectrometry. Both Ammons[66] and Knechtel
and Fraser[67] have digested sewage sludge with aqua regia to
dissolve trace metals. Carrondo et al[68] used a mixture of
nitric and sulfuric acid for this purpose. Tabatabai and
Frankenberger[69] digested sewage sludge with a nitric acid-
perchloric acid mixture and diluted the resulting solutions
with hydrochloric acid before measuring trace metal levels.
Martin et al[70] used nitric acid and hydrogen peroxide to decompose sewage sludge.

3.1.1.5. Plant Tissue Samples   Several acid digestion pro-
cedures have been reported for plant materials.[71-73] Most of
these are based on mixtures of nitric acid, sulfuric acid, and
perchloric acid. Price[74] recommends the following procedure
for the decomposition of plant material prior to the determination of trace metals by atomic absorption spectrometry:

> To 200 mg of finely ground plant material in a 100 mL
> Kjeldahl flask, add 0.5 mL of sulfuric acid, 1.0 mL
> of perchloric acid and 5 mL of nitric acid. Allow
> the digestion to proceed slowly at first and then
> increase the heat until the sulfuric acid refluxes
> down the side of the flask. When destruction of
> organic matter is complete, cool, transfer to a
> 50 mL calibrated flask and make up to volume with
> water.

Mutsch et al[75] have carried out similar digestions using 2-3 g
of dry plant material and a mixture of 25 mL of nitric acid
and 5 mL of perchloric acid. George and Kureisky[76] have used

the nitric-perchloric acid mixture to digest marine plant samples. Knause and Katz[22] were able to digest 1 g samples of pulverized leaf material with 10 mL of nitric acid in Kjeldahl flasks. They recommend allowing the material to remain in contact with the acid at room temperature overnight to reduce frothing.

3.1.1.6. Samples of Animal Tissues and Body Fluids   A wide variety of digestion procedures has been used to prepare animal tissues for atomic absorption spectrometry.

Body fluids, in some cases, require only dilution with a suitable solvent prior to aspiration into the spectrometer. Urinary calcium and magnesium levels are routinely determined after a 50-fold dilution of the sample with 0.5% lanthanum chloride solution.[77, 78]   Serum electrolytes (sodium and potassium) are measured after a 1:100 dilution with distilled, deionized water.[79] Marmar et al[80] have determined zinc in semen and expressed prostatic fluid after diluting 0.1-mL samples to 10 mL with 0.1% sodium chloride solution. Zinc standards were prepared with 0.1% sodium chloride solution to match the matrix of the samples. On occasion, denatured protein material would clog the capillary and interfere with the aspiration of the solution.

The high protein content of blood serum (6-8%) precludes the direct determination of trace metals by conventional flame atomic absorption spectrometry. Serious matrix effects are encountered. Extensive dilution of the samples is not possible because the metals are normally present at low concentrations, ie, Cu = 1.0-1.5 ppm, Fe = 1.0-1.5 ppm, and Zn = 0.5-2.0 ppm.[81] Modest dilution, 1:5 or 1:10, of the serum with 6% aqueous butanol has been recommended to prepare samples for the determination of iron, copper, and zinc.[79] The standards were prepared in 6% aqueous butanol and contained 140 mEq sodium per liter and 5 mEq potassium per liter.

Olson and Hamlin[82] recommend deproteinizing the blood serum with trichloroacetic acid prior to the determination of copper and zinc by atomic absorption spectrometry:

> Pipette 1 ml of serum and 1 ml of 10% trichloroacetic acid into a dry centrifuge tube. Mix carefully and centrifuge at 3,000 rev/min for ten minutes. Aspirate the supernatant into a lean air acetylene or

Table 3.1. Preparation of Blood Samples for Lead Determination
by Furnace Atomic Absorption Spectrometry

Preparation procedure	Reference
2 mL blood + 2 mL conc. nitric acid; incubate 30 min at 90°C; inject 2-µL samples	84
50 µL blood + 200 µL 0.1% aqueous Triton X-100; mix well; inject 15-µL samples.	85
1 mL blood + 9 mL 0.1% aqueous Triton X-100; mix thoroughly; inject 10-µL samples.	86
1 mL blood + 1 mL perchloric acid trichloroacetic acid mixture; incubate at room temperature for 1 h; dilute with 8 mL 0.01 M hydrochloric acid; inject 20-µL samples.	87
1 vol blood + 9 vol distilled water; mix gently; inject 10-µL sample followed by 10 µL 1:1 ammonia.	88
5 mL distilled water + 100 µL blood + 200 µL nitric acid; vortex or mix vigorously; centrifuge; autosample.	89

air propane flame. Prepare standards containing
0.1 mg/l of copper and zinc in 5% trichloroacetic
acid. Multiply the results obtained by the dilution
factor of 2.

For the determination of blood lead levels, acid digestion
followed by concentration was required for flame atomic absorption spectrometry.[83] The greater sensitivity of electrothermal atomization allows direct lead measurements on de-
proteinated and hemolyzed samples. Some of the procedures are
listed in Table 3.1.

3.1.1.7. Samples of Animal Tissues   Animal tissues are
frequently decomposed with a mixture of sulfuric acid, nitric
acid, and perchloric acid. Mixtures of nitric acid and sulfuric acid or nitric acid and perchloric acid are also

Table 3.2. Acid Digestion Procedures for Animal Tissues

Tissue	Digestion system	Reference
Foods	Sulfuric-nitric acid mixture	90
Foods	Sulfuric acid-hydrogen peroxide mixture	91
Beef	Nitric-perchloric acid mixture	92
Fish	Nitric acid in sealed vessel	93
Heart, lung, spleen, liver, kidney	Nitric-perchloric acid mixture	94
Kidney stones	Hydrochloric acid	95
Human liver	Sulfuric acid-hydrogen peroxide mixture	96
Bovine liver	Nitric acid and perchloric acid in sealed vessel	97
Bone	Nitric acid	98
Teeth	Perchloric acid	99
Hair	Nitric-perchloric acid mixture	100

employed for this purpose. Sulfuric acid-hydrogen peroxide mixture and nitric acid-hydrogen peroxide mixture are efficient digestion agents for animal tissues, and aqua regia is able to decompose these materials. Some applications of these agents are listed in Table 3.2.

A typical acid digestion procedure applicable to most tissues is:

> To a 250 or 500 mg sample contained in a Kjeldahl flask, add 10 mL of redistilled or high purity nitric acid. Allow the mixture to stand at room temperature for 2 hours. Heat to just below the boiling point for

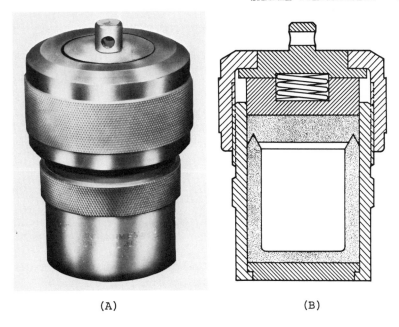

(A)   (B)

Figure 3.1. Parr Model 4745 acid digestion bomb.
Reproduced with permission from Parr
Instrument Company.

1 hour. Cool, and add 1 mL of high purity perchloric acid. Heat gently for 1 hour. Cool, and add
5 mL of high purity water. Heat until dense white
fumes are evolved. Cool and transfer to a 10 mL
volumetric flask.

3.1.1.8. High-Pressure Acid Decomposition Vessels   Acid
digestions are conveniently carried out in high-pressure decomposition vessels. Among the suppliers of these Teflon®-
lined steel cylinders are Parr Instrument Co. (Figure 3.1),
Perkin-Elmer Corp. (Figure 3.2), and Uni Seal Decomposition
Vessels, Ltd. Bernas[101, 102] has described many applications
of such vessels, and Katz et al have made use of these
devices for the acid decomposition of paper,[103] sewage
sludge,[104] and human hair.[105] For those digestions, 250 mg
of dry sample were weighed into the Teflon® container and
treated with 2.5 mL of redistilled nitric acid. The Teflon®
container was secured in the steel cylinder, and the assembly

Figure 3.2. Autoclave-2 for the digestion of difficultly
dissolved samples under pressure at increased
temperature in a closed PTFE vessel.
Courtesy of Perkins-Elmer.

was placed in a 100°C oven overnight. Complete digestion was
accomplished under these conditions.

Animal tissues have been decomposed by treatment with aqueous and nonaqueous solutions of tetraalkylammonium hydroxides.
Petering et al[106] found 10% aqueous tetramethylammonium hydroxide was as efficient as nitric acid for the digestion of liver
and kidney samples. Julshamn and Andersen[107] have recommended
"Lumatom" (a tetraalkylammonium hydroxide in toulene) for the
digestion of muscle biopsies. Cadmium, copper, and manganese
levels of bovine liver standard reference material samples
digested with this reagent agreed well with the results obtained
from samples of this material digested in nitric acid.

3.1.2. Dry Ashing and Fusion Techniques

Ashing and fusion have found frequent application in the
preparation of samples for atomic absorption spectrometry.
These procedures have been used in conjunction with a variety
of ashing aids and fluxes to destroy the sample matrix. The
residues are dissolved in appropriate solvents for introduction into the spectrometer.

3.1.2.1. Air Samples   Decomposition of filters used to
collect airborne particulates is rarely accomplished by ashing or fusion. Hoschler et al,[108] however, have ashed cellulose filters in platinum crucibles, fused the ash with sodium
carbonate, and dissolved the melt in hydrochloric acid.
Gallorini et al[109] used a sodium hydroxide-sodium peroxide
fusion as the first step in dissolving fly ash samples. Van
Loon[5] (pp. 252-253) recommends a procedure in which glass
fiber filters are ashed for 2 h at 500°C to destroy organic
matter and the residues are dissolved in hydrochloric acid-
hydrofluoric acid mixture.

3.1.2.2. Samples of Soils, Sediments, and Sludges   Water
and wastewater samples are not prepared for atomic absorption
spectrometry by ashing or fusion. Sewage sludge samples,
however, have been decomposed by these techniques. Bergman
et al[110] have employed dry ashing at 500°C followed by
leaching with 3 N hydrochloric acid. Ritter et al[111] have
reported a similar procedure for the preparation of sewage
sludge prior to atomic absorption spectrometry:

>One g samples of sludge were weighed and transferred
>into acid-washed porcelain crucibles. The samples
>were ignited at 550°C for 2 1/2 hours in a muffle
>furnace. They were then cooled and placed in acid-
>washed, 50 ml Folin-Wu test tubes. Twenty five ml
>of 3N hydrochloric acid were added to each tube,
>and the tubes were placed in holes in an aluminum
>block heater. The samples were heated at 120°C for
>2 hours, after which they were filtered through
>Whatman #42 paper and diluted to volume with de-
>ionized water in 50 ml volumetric flasks.

Soils and sediments are decomposed by fusion procedures.
Brown and Newman[112] have decomposed soils by fusing the
samples with lithium metaborate and dissolving the melts with
nitric acid. Van Loon and Parissis[113] developed this method
in which a 0.2- to 0.5-g sample is mixed with 2 g of lithium
metaborate in a platinum crucible and heated to 900°C for
15 min. The cooled melt is then placed in a beaker containing

150 mL of water and 8 mL of nitric acid and stirred until
solution is complete. Nadkarni and Morrison[114] have used a
sodium peroxide fusion to decompose lake sediments, and Van
Loon[5] has described a similar procedure for the decomposition
of other geological materials.

Direct ignition of the sample followed by acid leaching of
the ash is employed to destroy organic material. Motuzova
and Obukhov[115] have calcined soil samples at 450°C for 2 h
and leached the ashes with sulfuric acid-nitric acid mixture
for the determination of copper, manganese, and zinc.

3.1.2.3. Plant Tissue Samples

Dry ashing is frequently
employed for the decomposition of plant tissues. Jones[116]
recommends the use of this technique if the following precautions are observed:

1. The ashing temperature should not exceed 500°C.

2. The ashing vessels should be suspended in the muffle
   furnace to avoid superheating when the base is in contact with the heating element in the floor of the
   furnace.

3. Ashing aids such as magnesium oxide, which can efficiently trap and retain a number of volatile elements
   during dry ashing, may need to be avoided.

4. High-walled ashing vessels should be used rather than
   open, flat-type ashing vessels.

Garten et al[117] have used dry ashing to prepare the component
parts of different plant species from the floodplain community
along a coastal plain stream in South Carolina. Gorshkov
et al[118] prepared plant material by ashing the sample and
leaching the ash with hydrochloric acid. Plant material was
dry ashed using nitric acid combined with either potassium
bisulfate or sulfuric acid as ashing aids by Heanes.[119]
Price[74] recommends the following dry ashing procedure for
plant material:

> Place 200 mg of finely ground sample, contained in a
> silica dish or platinum dish, in a cool muffle
> furnace. Allow the temperature to rise to 450°,

and ash at this temperature for 3 hours. Cool, dissolve the residue in 5 ml of 5 M hydrochloric acid.
Add a few drops of nitric acid and evaporate to dryness on a water bath. Redissolve in 5 ml of 5 M
hydrochloric acid, warm, filter into a 50 ml calibrated flask and make up to volume with water.

3.1.2.4. Animal Tissue Samples   Ashing and fusion procedures are not normally employed to prepare samples of body
fluids for atomic absorption spectrometry. Nonetheless,
Cornelis et al[120] ashed lyophylized urine samples at 450°C
and dissolved the residues in nitric acid to prepare the material for the determination of molybdenum. Wang[121] has used
the following procedure to prepare tissue from the lungs for
cobalt determinations:

    Tissue samples were dried at 110° and powdered. Be-
        tween 10 and 100 mg of powder were weighed into
        platinum crucibles and ashed at 500°-550° for 4-5
        hours. The ashes were leached with 1 ml of concen-
        trated hydrochloric acid to dissolve the cobalt
        compounds.

The following procedure was used by Locke[96] to prepare human
liver tissue for atomic absorption spectrometry:

    Drained, fresh human livers were cut into small por-
        tions. Into Vitresil crucibles were weighed 14 g
        samples of chopped liver. The crucibles with their
        samples were placed in a 500° muffle furnace over-
        night and then for 2 hours more at 550°. The sam-
        ple ashes (200 mg) were dissolved in 1 ml of con-
        centrated sulfuric acid by heating on a steam bath
        for 1 hour. A few drops of 30% hydrogen peroxide
        were added and the heating was continued for 1 hour.
        The mixture was diluted to 20 ml with water.

3.1.2.5. Techniques Involving Oxygen Oxidation Devices
The Schönenger flask and the Parr bomb have found modest
utilization in preparing plant and animal tissue samples for
atomic absorption spectrometry. Lidums[122] has employed this
technique for the determination of mercury in fish tissue by

cold vapor atomic absorption spectrometry. Since the combustion took place in a closed system, loss by volatilization was minimal.

The low-temperature oxygen plasma asher appears to have much to offer for the decomposition of tissue samples. Oxygen radicals are produced by a radiofrequency discharge. These radicals destroy the organic matrix of the samples. Locke[96] has ashed human liver samples in such a device at a temperature well below 100°C. Although recoveries of magnesium, calcium, manganese, copper, iron, zinc, rubidium, and cadmium compared favorably with those obtained by conventional wet digestion and dry ashing procedures, the low-temperature oxygen plasma ashing required some 60 h to ash 14 g of fresh liver sample. Zief and Mitchell[33] had previously cited good recoveries for most metals but identified losses of silver and mercury from blood samples subjected to low-temperature oxygen plasma ashing.

3.2. Concentration and Separation Methods

Samples of water and waste water or the solutions resulting from the decomposition of solid samples sometimes require further treatment to raise the levels of analyte to the optimal concentration ranges for atomic absorption spectrometry or to remove interfering elements from the solution containing the analyte. The common treatments for such procedures are:

1. Solvent extraction. This technique is probably the most common for concentrating the analyte or removing interferences.

2. Evaporation of solvent. Although straightforward, this approach is vulnerable to losses by volatilization, and it serves to increase the levels of both analyte and interfering elements.

3. Coprecipitation. Redissolving the precipitate and the coprecipitated analyte gives both increases in concentration and a better defined matrix.

4. Ion exchange chromatography. This technique gives significant increases in the concentration of analyte, and it efficiently removes some interfering species especially when chelating resins are used.

5. Electrodesposition. This technique has been used to isolate and concentrate analyte from complex systems.

3.2.1. Solvent Extraction

Solvent extraction enjoys a favored position among these
techniques because of its speed, simplicity, and broad scope.
Extractions require only a few minutes to perform, and apparatus no more complicated than a separatory funnel is employed. Extraction procedures are applicable to both trace
and macro quantities.

Solubility in organic solvents is not a normal characteristic of simple metal salts. Hence, their extraction from
aqueous media requires the replacement of the water molecules
hydrating the metal ion by some other coordinating group and
the neutralization of charges by association with other ionic
species. The factors involved in the extraction of metal ions
from aqueous solutions into immiscible organic solvents have
been reviewed by Katz.[123, 124]

The formation of an extractable species frequently involves
the formation of a metal chelate. The Lewis acidity of the
metal ion, the Lewis basicity of the polydentate ligand, and
the pH of the aqueous solution are important factors in the
formation of such a compound. In the general case, the formation of the metal chelate, ML_2, is governed by

$$K_f = (ML_2)/(M^{+2}) \ (L^-)^2$$

The polydentate ligand, L^-, is the conjugate base of the weak
acid HL and (L^-) is dependent upon K_a of the acid and the pH
of the system. Pyrrolidine dithiocarbamic acid* forms chelate
compounds with some two dozen metal ions. The pH ranges for
their formation and extraction are listed in Table 3.3.

The extraction of the metal chelate is dependent upon its
relative solubility in the extracting organic phase and in
the initial aqueous phase, ie,

---

*The ammonium salt ammonium pyrrolidine dithiocarbamate (APDC)
is commonly used.

Table 3.3. pH Ranges for the Formation of Metal-APDC
Chelates and Their Extraction into MIBK[a][80, 125]

Metal ion	Formation range	Extraction range
Antimony (III)	2-9, 2-9	1, 2-4
Arsenic (III)	2-6, 0-6	2-6, 0-4
Bismuth (III)	2-14, 0-14	1-10, 1-6, 1-10
Cadmium (II)	2-14, 0-14	1-10, 1-6, 0-11
Chromium (?)	2-6, 2-9	3-9, 3-7
Cobalt (II)	2-14, 1-14	1-10, 2-4, 1-10
Copper (II)	2-14, 0-14	1-10, 1-8, 0-14
Iron (II)	2-14	—
Iron (III)	2-14	2-5
Lead (II)	2-14, 0-14	1-10, 1-6, 0-8
Manganese (II)	2-14, 2-12	5-10, 2-4, 4-6
Mercury (II)	2-14, 0-14	1-10, 0-10
Nickel (II)	2-14, 1-14	1-10, 2-4, 1-10
Selenium (?)	2-9, 2-10	3-6, 3-6
Silver (I)	2-14, 0-14	1-10, 0-14
Tellurium (?)	2-6, 2-6	3-5
Thallium (?)	2-14, 1-14	3-10, 2-12
Zinc (II)	2-14, 1-14	1-10, 2-6, 1-10

[a]MIBK, methyl isobutyl ketone.

$$D = (ML_2)_o / (ML_2)_a$$

Efficient extractions demand favorable K_f and favorable D values.

The extracting organic phase must possess the following characteristics:

1. Immiscibility with the aqueous phase

2. High solvency for the metal chelate

3. Combustion properties compatible with the atomization process

Methyl isobutyl ketone (MIBK) meets these requirements
quite well. It is an efficient extractant for ammonium
pyrrolidine dithiocarbomate (APDC) chelates and it routinely
allows 10-fold concentration increases. In addition, the
atomization of the metal chelate is more efficient in MIBK
than it is in aqueous media. This leads to a two- to fivefold
enhancement in the absorbance signal. Hence, extraction of
metal-APDC chelates from 100 mL of aqueous solution into 10 mL
of MIBK and aspiration of the organic phase frequently increases sensitivity by 30 to 40 times. The procedure required
by the U.S. EPA[14] is as follows:

> Special Extraction Procedure: When the concentration of
> the metal is not sufficiently high to determine
> directly, or when considerable dissolved solids are
> present in the sample, certain metals may be che-
> lated and extracted with organic solvents. Ammonium
> pyrrolidine dithiocarbamate (APDC) in methyl iso-
> butyl ketone (MIBK) is widely used for this purpose
> and is particularly useful for zinc, cadmium, iron,
> manganese, copper, silver, lead and hexavalent
> chromium. Tri-valent chromium does not react with
> APDC unless it has first been converted to the hexa-
> valent form. Aluminum, beryllium, barium, and
> strontium also do not react with APDC. While the
> APDC-MIBK chelating-solvent system can be used
> satisfactorily, it is possible to experience
> difficulties.
>
> 1. Transfer 200 ml of sample into a 250 ml sep-
>   aratory funnel, add 2 drops of bromo-phenol
>   blue indicator solution and mix.
>
> 2. Prepare a blank and sufficient standards in
>   the same manner and adjust the volume of each
>   to approximately 200 ml with deionized dis-
>   stilled water. All of the metals to be deter-
>   mined may be combined into a single solution
>   at the appropriate concentration levels.
>
> 3. Adjust the pH by addition of 2 N ammonia solu-
>   tion until a blue color persists. Add HCl

dropwise until the blue color just disappears; then add 2.0 ml HCl in excess. The pH at this point should be 2.3.

4. Add 5 ml of PDCA-chloroform reagent* and shake vigorously for 2 minutes. Allow the phases to separate and drain the chloroform layer into a 100 ml beaker.

5. Add a second portion of 5 ml PDCA-chloroform reagent and shake vigorously for 2 minutes. Allow the phases to separate and combine the chloroform phase with that obtained in step 4.

6. Determine the pH of the aqueous phase and adjust to 4.5.

7. Repeat step 4 again combining the solvent extracts.

8. Readjust the pH to 5.5 and extract a fourth time. Combine all extracts and evaporate to dryness on a steam bath.

9. Hold the beaker at a 45° angle, and slowly add 2 ml of concentrated distilled nitric acid, rotating the beaker to effect thorough contact of the acid with the residue.

10. Place the beaker on a low temperature hot plate or steam bath and evaporate just to dryness.

11. Add 2 ml of nitric acid (1:1) to the beaker and heat for 1 minute. Cool, quantitatively transfer the solution to a 10 ml volumetric flask and bring to volume with distilled water. The sample is now ready for analysis.

---

*Pyrrolidine dithiocarbamic acid (PDCA): Prepare by adding 18 ml of analytical reagent grade pyrrolidine to 500 ml of chloroform in a liter flask. Cool and add 15 ml of carbon disulfide in small portions and with swirling, dilute to 1 liter with chloroform. The solution can be used for months if stored in a brown bottle in a refrigerator.

Price[74] proposes the extraction of the APDC metal chelates into MIBK and aspiration of the organic phase:

> To 50 ml of sample, add 5 ml of 1% aqueous APDC solution
> and adjust the pH to the optimum value from Table
> 3.3. Transfer the solution to a 100 ml separatory
> funnel, extract the complex into 4 ml of MIBK by
> vigorously shaking for 30 seconds, then stand for
> two minutes. Transfer the aqueous phase to another
> separatory funnel and repeat the extraction with
> 1 ml of MIBK. Discard the aqueous phase, combine
> the extracts in the first funnel, mix and filter
> through a cotton wool plug into a small beaker. The
> extract is aspirated into the atomic absorption
> spectrometer.

The Perkin-Elmer "cookbook"[7] contains a similar method for the determination of trace metals in natural waters. Diethyldithiocarbamate (DDC) rather than APDC is the chelating agent used in this procedure.

Brooks et al[126] applied the APDC-MIBK extraction system to the determination of cobalt, copper, iron, lead, nickel, and zinc in sea water. They determined distribution coefficients of 600, 600, 300, 500, 350, and 100, respectively, for these metal chelates extracted from sea water in the pH range of 3 to 5. Using 750-mL samples of sea water and 20 mL of MIBK, the concentrations of these elements in the organic phase were increased almost 40 times their original levels in the aqueous phase. These increases in concentration coupled with the more efficient atomization achieved by aspiration of the organic phase allowed detection limits of 0.1 ppb.

Meranger et al[127, 128] employed the extraction of APDC chelates into MIBK for the determination of some of the metals in their national survey of Canadian drinking water samples, and Katz[129] has used this technique for the determination of lead in his survey of potable water supplies in southern New Jersey.

Lead has been extracted from blood and urine as the APDC chelate into MIBK prior to determination by atomic absorption spectrometry.[7, 83, 130] A typical procedure[7] is as follows:

> Pipette 5 ml of heparinized whole blood into a graduated
> 10 ml stoppered centrifuge tube. Centrifuge for

20 minutes at 3000 rev/minute then remove and discard the plasma. Note the volume of the packed red
cells. Add 1 drop of saponin/triton solution (5 ml
of Triton X plus 5 g of saponin in 25 ml of water)
and mix. Into the mixture pipette first 2 ml of 2%
APDC solution, shake for 30 seconds, and then add
3.0 ml of MIBK. Stopper the test tube, shake for
1 minute then centrifuge at 3000 rev/min for 5 minutes. Aspirate the top layer into a lean air acetylene flame. Prepare standards by taking 2 ml of lead
standard solutions containing 0.2 µg of lead per ml
through the procedure. Read the lead content of the
sample extract in µg, then lead content of red cells
in µg/100 ml = (µg in extract × 100)/v where v is the
volume of the red cells.

3.2.2. Evaporation of Solvent

Although an obvious approach to concentrating analyte, the
evaporation of solvent suffers two major disadvantages. If
the original sample has a high total dissolved solids (TDS)
content, the TDS of the concentrated sample may be so high as
to cause interferences in the atomic absorption measurements.
It is also possible to experience losses of analyte through
precipitation or coprecipitation if the solubility of some
component of the system is exceeded during the evaporation
process. It is equally possible to experience losses by
volatilization during the evaporation process. Of lesser significance is the slowness of the volatilization process: Several hours are required to achieve a 10-fold concentration
increase by evaporating a liter sample to 100 mL.

In spite of these drawbacks, the U.S. EPA[9] does permit
utilization of this technique. The digestion procedure used
for the determination of total metals (see Section 5.2.1) can
be used to concentrate drinking water samples: "the final
volume may be reduced to effect up to a 10 × concentration of
the sample, provided the total dissolved solids in the original
sample do not exceed 500 mg/L, the determination is corrected
for any non-specific absorbance, and there is no precipitation
loss."

3.2.3. Coprecipitation

Coprecipitation offers 10- to 50-fold concentration in-creases. The technique, however, has several disadvantages:

1. The precipitation process is lengthy and tedious.

2. The precipitating agent must be of high purity to pre-vent contamination.

3. If the carrier precipitate (scavenging agent) cannot be destroyed during the redissolving process, the final solution will have a high TDS and be subject to the same interferences encountered when solutions are concen-trated by evaporation of solvent.

Bowen[131] has described the efficiency of coprecipitation with scavenging precipitates such as ferric hydroxide. This agent has been used to scavenge trace metals from sea water,[125] and bismuth hydroxide has served a similar function in col-lecting lead from blood and urine.[79]

Mallory[132] has developed a thioacetamide precipitation pro-cedure for determining trace elements in water. Tin(II) sul-fide was used as the scavenging agent for antimony, bismuth, cadmium, copper, and lead, and indium sulfide was the carrier precipitate for beryllium, chromium, iron, titanium, and zinc. The tin(II) was precipitated with hydrogen sulfide liberated by the hydrolysis of thioacetamide in acid medium. The sul-fide for the precipitation of indium resulted from the hydroly-sis of thioacetamide in alkaline medium.

Thioanalide (2-mercapto-N-naphthyl-acetamide) has been used as a carrier precipitate.[131,123] Recoveries in excess of 95% were obtained from 30% aqueous ethanol solution at pH 10 for cobalt, chromium, hafnium, mercury, indium, iridium, manganese, osmium, ruthenium, tin, titanium, and zinc.

3.2.4. Ion Exchange Chromatography

Ion exchange techniques are frequently employed for the separation and/or concentration of analyte. Using traditional strong acid or strong base resins, cationic species such as calcium can be separated from anionic interferences such as phosphate. These resins have also been used to concentrate trace metals from dilute solutions. Ordinary ion exchange resins are, however, of limited use in concentrating trace elements from sea water, urine, or blood plasma.

Chelex 100* is a styrene-divinylbenzene copolymer containing iminodiacetate functional groups:

It differs from conventional strong acid resins in several respects:

1. The selectivity of Chelex 100 is a function of the iminodiacetate functional group rather than ionic size or charge.

2. The bond strength of Chelex 100 is almost 10 times greater than that of conventional strong acid resins.

3. Chelex 100 demonstrates slower exchange kinetics than do other types of ion exchangers.

Chelex 100 demonstrates a very strong attraction for transition metal ions even from solutions of high salt concentrations. Chelex 100 has been used to concentrate cadmium, copper, nickel, and zinc from sea water prior to their determination by atomic absorption spectrometry.[5]

> Filter sea water samples through a 0.5 μm membrane filter and allow 10 liter aliquots to pass through the column of Chelex 100; the flow rate should not exceed 300 ml/hr. If only zinc, copper and nickel are to be determined, 1 liter aliquots are sufficient. Wash the column with 250 ml of water and reject the washings. Elute copper, nickel, zinc and cadmium

---

*Purified form of Dowex A-1 available from Bio Rad Labs, Richmond, CA.

with 30 ml of 2 N nitric acid and then elute cobalt
with 20 ml of 2 N hydrochloric acid. Place the
eluate in a 50 ml silica conical flask, cover with a
silica bubble stopper, and evaporate to dryness on a
hot plate at low temperature. Add to the flask, by
means of a pipette, 1 ml of 0.1 N nicric acid and
when the residue has dissolved add 9 ml of acetone.
Determine copper, nickel, cadmium and cobalt in the
appropriate solution using the atomic absorption
spectrometer and the resonance lines at wavelengths
3247, 2318, 2287 and 2406 Å, respectively. Dilute
1 ml of the acetone solution from the 2 N nitric
acid eluate to 5 ml with 9% aqueous acetone and use
the resultant solution for the determination of zinc
using the 2137 Å zinc resonance line. Determine the
reagent blank for the method in the same manner
using sea water that has been stripped of trace elements by passage through a column of Chelex 100.
Calibrate the instrument in the range of 0.5-2.5 µg/
ml with standard solutions of the elements in 90%
acetone; 90% acetone is used as a blank for the
calibration.

Mygaard and Hill[133] have compared treatment with Chelex 100
resin to APDC-MIBK extraction for the concentration of copper,
zinc, cadmium, and lead from sea water. Neither procedure
could be identified as superior to the other for estimation
of biologically available metal.

Resins other than Chelex 100 have found application in the
separation and concentration of trace metals prior to their
determination by atomic absorption spectrometry. Korkisch[134]
used anion exchange chromatography on Dowex 1 to concentrate
cadmium, cobalt, copper, manganese, lead, and zinc from the
nearshore waters of the Adriatic Sea. Barnes and Genna[135]
have described the preparation of a poly (dithiocarbamate)
resin, and its application to the determination of trace
metals in urine. Concentration factors of 125 were achieved
with this chelating resin, and recoveries were better than 95%
for cadmium, copper, lead, mercury, nickel, selenium, and tin.
Enrichments of 100-fold for trace metals from fresh water and
from sea water were obtained by Burda et al.[136] They employed
a cellulose exchanger containing 1-(2-hydroxyphenylazo)-

2-naphthol to concentrate cadmium, cobalt, chromium, copper, iron, mercury, manganese, molybdenium, nickel, lead, and zinc from 5-L water samples. Elution was with 50 mL of 1 M hydrochloric acid.

3.2.5. Electrodeposition

While popular for isolating radioactive materials, electrodeposition has not found wide use as a separation and concentration technique in atomic absorption procedures. On the basis of fundamental electrochemical considerations, it appears that this technique should allow the selective electrodeposition of trace elements from a large volume of a solution containing a high concentration of alkali and alkaline earth metal salts. Subsequent solubilizations of these elements from the electrode should give significant enrichments in a solution of defined matrix. Kinetic factors for such a deposition, however, are frequently unfavorable, and uncertainties about the matrix of the original solution lead to ill-defined decomposition potentials. Nonetheless, Hernandez-Mendez et al[137] have determined gold by conventional flame atomic absorption after electrochemical preconcentration on a spiral platinum electrode. At an applied potential of -300 mV (vs. saturated calomel electrode (SCE)), 50 ml of solution could be electrolyzed in 30 min. The detection limit, under these conditions, was 1.6 ng/mL.

3.3. Considerations for Contamination and Loss of Analyte

Samples are subject to contamination or loss of analyte at essentially any point in the collection-analysis process. Hence, special precautions are taken in cleaning sampling devices and sample containers, in collecting the samples, and in insuring their integrity through the addition of specific preserving agents and the placement of limitations on holding times and conditions. Similar special precautions also apply to the routine laboratory operations directed to sample preparation. Sansoni and Iyengar[138] and Iyengar and Sansoni[139] have recently reviewed many of the problems involved in sampling, storing, and preparing biological materials for trace element analysis.

3.3.1. Purity of Reagents

Recognizing that any reagent added for the preparation of the sample is a potential source of contamination, dry ashing would appear to be superior to acid decomposition. Table 3.4 lists some of the impurities of three commercially available laboratory grades of nitric acid.[140] The digestion of 200-mg samples of fresh human liver with 10 mL of each of these acids

Table 3.4. Impurities in Nitric Acid

	#4801 ultrex (ppb)	#9598 instra-analyzed (ppb)	#9601 analyzed reagent (ppb)
Aluminum	2	50	
Arsenic	<0.001	5	5
Barium	<10	50	
Beryllium	<0.02		
Bismuth	<0.1		
Cadmium	<1	5	
Calcium	2	50	
Chromium	0.3	50	100
Cobalt	<0.1	5	
Copper	0.07	5	50
Gold	<1		
Iron	0.5	20	200
Lead	0.5	5	
Magnesium	0.3	50	
Manganese	0.5	5	
Mercury	<0.1	5	
Nickel	0.2	5	50
Potassium	<5	50	
Silver	<0.05	10	
Sodium	30	500	
Strontium	<1	50	
Tin	<0.5	50	
Titanium	0.2		
Zinc	<1	10	

would give solutions containing trace metals from both the
liver sample and from the acids. The chromium and nickel
contributions from the acids would be:

	Chromium (μg)	Nickel (μg)
Ultrex	0.003	0.002
Instra-analyzed	0.5	0.05
Analyzed reagent	1	0.5

The chromium and nickel levels of human liver are on the order
of 0.1 ppm.[81] The contributions of chromium and nickel from
the liver sample would be only 0.02 μg. Obviously, the contributions of chromium and nickel form the instra-analyzed
and the analyzed reagent grades of acid exceed the 0.02 μg of
chromium and the 0.02 μg of nickel from the liver sample.
Only with the ultrex grade acid is the reagent blank sufficiently low to allow measurement of these materials in liver.

Successful trace analysis demands not only high-purity
acids but also high-purity water. Several commercially available units are able to supply "reagent water" suitable for
trace analysis.[141-144] The American Society for Testing and
Materials (ASTM)[15] identifies four categories of reagent water:

TYPE I grade of reagent water shall be prepared by the
    distillation of feed water having a maximum con-
    ductivity of 20 μMHOS/cm at 25° followed by polish-
    ing with a mixed bed of ion exchange material and a
    0.2 μm membrane filter.

TYPE II grade of reagent water shall be prepared
    by distillation using a still designed to produce a
    distillate having a conductivity of less than
    1.0 μMOHS/cm at 25°. Ion exchange, distillation or
    reverse osmosis may be required as an initial treat-
    ment prior to distillation if the purity cannot be
    obtained by a single distillation.

TYPE III grade of reagent water shall be prepared
    by distillation, ion exchange, reverse osmosis or a
    combination thereof followed by polishing with a
    0.45 μm membrane filter.

Table 3.5. Specification for Reagent Water

	Type I	Type II	Type III	Type IV
Total matter (mg/L)	0.1	0.1	1.0	2.0
Conductivity at 25°C (μmho/cm)	0.06	1.0	1.0	5.0
Resistivity at 25°C (Mohm-cm)	16.67	1.0	1.0	0.2
pH at 25°C	[a]	[a]	6.2-7.5	5.0-8.0
Permanganate color retention (min)	60	60	10	10
Soluble silica (mg/L)	ND[b]	ND[b]	10	NL[c]

[a] Electrodes contaminate system; test is meaningless.

[b] ND, not detected.

[c] NL, no limit.

TYPE IV grade of reagent water may be prepared by distillation, ion exchange, reverse osmosis or a combination thereof.

The specifications for each type of reagent water are summarized in Table 3.5.

Ultrapure reagents are available from several manufacturers.[140, 145-147] Zief and Mitchell[33] have described procedures for the preparation of such high-purity materials. Hydrochloric acid, nitric acid, perchloric acid, sulfuric acid, and hydrofluoric acid are prepared by subboiling distillation, isoprestic distillation, and/or gaseous saturation of high-purity water. Isoprestic distillation of reagent grade ammonia is also used to prepare this material in a high-purity form. Recrystallization and extraction are employed for the preparation of high-purity salts, buffers, and complexing agents.

Contamination by impurities in reagents is easily identified with the reagent blank. When the reagent blank is sufficiently low, less than 10% of the values measured for the samples, a correction may be applied. If the blank values are of the same order of magnitude as that determined for the samples, purification of reagents is needed.

3.3.2. Cleaning of Laboratory Glassware

A second source of contamination is improperly cleaned laboratory glassware. The U.S. EPA has included the following procedure in its "General Requirements for Metals Analysis"[14] for cleaning glassware, plastic containers, and sample tubes:

1. Thoroughly scrub with detergent and water.

2. Rinse with a solution of one part concentrated nitric acid to one part water.

3. Rinse with water.

4. Rinse with a solution of one part hydrochloric acid to one part water.

5. Rinse with water.

6. Rinse with deionized, distilled water.

7. Dry the plastics at 50°C, the glassware at 105°C.

3.3.3. Volatilization of Analyte

Behne[148] identified the absorption of analyte on the surfaces of containers and the volatilization of analyte from the containers as the major factors responsible for losses. While absorption losses to the walls of the sample container are retarded by the addition of appropriate preservatives, absorption and volatilization losses during heating of the sample are of serious concern. Bowen[131] has cited significant losses of mercury from aqueous solutions stored at 60°C. Similarly, he called attention to losses during the dry ashing of biological material. When heated in air at 450°C to 550°C, losses of arsenic, cadmium, gold, iron, lead, mercury, osmium, ruthenium, selenium, and silver occurred. In addition, he cited losses of gold, mercury, and silver during low-temperature oxygen plasma ashing. Christian and Feldman[149] have also reviewed volatilization losses during dry ashing. They have also identified absorption losses during dry ashing procedures. The oxides of some elements, copper in particular, undergo reduction to the metal and diffuse into the crucible when dry ashed in the presence of organic matter.

Ashing aids have been employed to reduce losses of analyte by volatilization and/or absorption. Magnesium nitrate has found extensive use in this capacity. Watling and Wardale[150] found significant (50%) increases in the lead levels of fish tissue and of algae when magnesium nitrate was used as an ashing aid compared to the same tissues prepared by conventional dry ashing. They also found 10- to 100-fold increases in the nickel content of oyster, fish, algae, heart, and brain tissue prepared with magnesium nitrate ashing aid over the levels of nickel in these tissues determined in samples prepared without the ashing aid. The increases in nickel levels, from 0.4 to 4 ppm in samples prepared without the ashing aid to 55 to 59 ppm in samples prepared with the magnesium nitrate, were due to contamination. It is necessary to establish blank values whenever ashing aids or fusion mixtures are used.

The loss of analyte, unlike the introduction of contamination, cannot be determined from the blank values. Recovery studies in which known amounts of analyte are carried through the dry ashing procedure are used to evaluate volatilization and absorption losses. In some cases, samples are split and spiked, and recoveries are calculated from the differences between the results obtained from the spiked and the unspiked split samples. Hislop and Williams[151] have confirmed the need to match the chemical forms of the analyte and the spiked in such recovery studies. They found, in their studies on the behavior of lead during the dry ashing of bone, that a lead carbonate spike was volatilized more readily than a lead phosphate spike. Behne and Brätter[152] found 100% recoveries of inorganic chromium carried through four different decomposition procedures. The chromium content of a yeast sample prepared by these four procedures, however, reflected volatilization losses. It is not unreasonable to expect that the spike is not biologically incorporated and behaves differently from the analyte under the experimental conditions.

3.3.4. Comparison of Acid Decomposition and Dry Ashing

While it appears that acid decomposition procedures are subject to contamination from reagents and dry ashing procedures suffer from volatilization losses, both procedures have been used successfully to prepare samples for atomic absorption spectrometry. With the availability of high-purity acids, the acid decomposition procedures enjoy a greater popularity.

Rice[153] has compared three acid decomposition procedures and lithium metaborate fusion for the determination of potassium in geological materials by atomic absorption spectrometry. Samples were decomposed by hydrofluoric acid-nitric acid digestion, in sealed polyethylene bottles, hydrofluoric acid-nitric acid digestion in platinum dishes, hydrofluoric acid-nitric acid-sulfuric acid digestion, and lithium metaborate fusion in platinum crucibles. Statistical analysis of the results indicates no significant difference between the sealed bottle digestion and the lithium metaborate fusion. The results obtained from the samples prepared by hydrofluoric acid-nitric acid digestion in platinum dishes and the digestion with hydrofluoric acid-nitric acid-sulfuric acid mixture were, however, significantly lower.

Articola-Fortuny and Fuller[154] have analyzed sewage sludge samples prepared by six different methods:

1. Wet oxidation: repeated treatment with hydrogen peroxide.

2. Wet oxidation: digestion in nitric acid-hydrogen peroxide mixture

3. Wet oxidation: digestion in nitric acid-perchloric acid mixture

4. Parr bomb: high pressure acid decomposition with aqua regia

5. Dry ashing: dry ashing with nitric acid ashing aid

6. Carbonate fusion: fusion with sodium carbonate

Their data show very poor recoveries from samples treated with only hydrogen peroxide. Wet oxidation with nitric acid and hydrogen peroxide appears to be less efficient than the re-

maining procedures. Methods 3, 4, 5, and 6 show similar
recoveries.

Articola-Fortuny and Fuller[155] have also applied these six
procedures to the preparation of manure samples. The samples
treated with hydrogen peroxide showed very low recoveries of
all metals. Digestion with nitric acid and hydrogen peroxide
and dry ashing showed, for most metals, lower recoveries than
obtained by carbonate fusion, high-pressure acid decomposition,
or digestion in nitric acid-perchloric acid mixture.

Delfino and Enderson[156] have compared three acid decomposition procedures—digestion in nitric acid, digestion in nitric
acid-hydrogen peroxide mixture, and digestion in aqua regia—
and two dry ashing procedures, muffle furnace ignition and
low temperature oxygen plasma ashing. Compared to the acid
decomposition procedures, the muffle furnace ignition was in-
ferior in terms of iron and chromium recoveries. All five
methods appeared equivalent in terms of cadmium, copper, lead,
manganese, and zinc recoveries.

Ritter et al[111] have also compared acid decomposition procedures to dry ashing: they concluded that dry ashing was
the best way to prepare sewage sludge samples for atomic absorption spectrometry. Compared to digestion in nitric acid
and digestion in aqua regia, dry ashing appeared to give
superior recoveries.

Carrondo et al,[68] on the basis of a comparison of acid de-
compositions and dry ashing of sewage sludge, have found that
samples prepared by ashing at 450°C show lower iron, magnesium,
and aluminum levels than samples prepared by digestion with
nitric acid-hydrogen peroxide mixtures or with nitric acid-
sulfuric acid mixtures. Samples digested in perchloric acid-
hydrofluoric acid-nitric acid mixtures show even higher levels
of aluminum and magnesium.

Katz et al[104] have reported that digestion in nitric acid-
hydrogen peroxide mixture is far superior to dry ashing for
preparing sewage sludge samples prior to atomic absorption
spectrometry. The results from samples prepared by high-
pressure acid decomposition (Parr bomb) agree well with those
from samples prepared by nitric acid-hydrogen peroxide digestion. Iron and nickel levels for samples prepared by dry
ashing were much lower than those for samples prepared by
acid decomposition.

Feinberg and Ducauze[157, 158] have evaluated dry ashing

with and without sulfuric acid ashing aid at temperatures of 450°, 650°, 850°, and 980°C. Some two dozen foods were prepared for lead and cadmium analysis by direct calcination at 750°C. In most cases, the recoveries were 95% or better when sulfuric acid ashing aid was used. No losses of cadmium or lead were observed as the ashing temperature was increased from 450° to 980°C.

Ross and Umland[159] have determined the levels of calcium, vanadium, and zinc in petroleum samples prepared by dry ashing, wet oxidation, Schöniger combustion, low-temperature oxygen plasma ashing, and high-pressure acid decomposition (Parr bomb). The best recoveries were obtained in samples prepared by high-pressure acid decomposition.

Locke[96] determined calcium, magnesium, iron, zinc, copper, manganese, rubidium, and cadmium in samples of liver tissue prepared by low-temperature oxygen plasma ashing, dry ashing at 500°C, and digestion in nitric acid-sulfuric acid mixture. Some loss of copper, attributed to retention by the crucibles, was noted in samples prepared by dry ashing. No other differences were observed between the various approaches to sample preparation.

Watson[160] has reported on an interlaboratory comparison of trace element levels in leaf materials. Some of the participating laboratories prepared their samples by dry ashing; others used acid decomposition. Iron and copper results tended to be higher when samples were prepared by acid decomposition.

It appears that there is no universal method to prepare all samples for atomic absorption spectrometry. Each method has its specific advantages and disadvantages for the various metals in the different sample matricies. It is necessary that laboratories whose data will be used to establish compliance monitoring use the same methods and procedures of sample preparation. Regulatory agencies have published such common procedures. Coupled with a rigid quality assurance program, these procedures should yield reliable data.

4. METHODS FOR COMPLIANCE AIR QUALITY MONITORING

The methods contained in this section are applicable to
compliance monitoring of air quality in the work place.
These methods are contained in more detail in the NIOSH Manual of Analytical Methods,[10] Volumes 1 through 6. They are a
necessary part of all standards which set limits for toxic
substances. Many of the methods have been recommended to
the Occupational Safety and Health Administration (OSHA) for
use in compliance monitoring, and many have been validated
using an improvement of the protocol developed in 1974 for
the joint National Institute of Occupational Safety and
Health (NIOSH)-OSHA Standards Completion Program.

4.1. General Procedure for Metals:
NIOSH Method P&CAM 173

This procedure describes a general method for the collection, dissolution, and determination of trace metals in industrial and ambient airborne material. The samples are collected on membrane filters. Nitric acid digestion is used to
destroy organic material and dissolve the metals present in
the sample. The determination of metal levels in the sample
is made by atomic absorption spectrometry.

For personal sampling, a 37-mm diameter, 0.8-μm pore size
cellulose ester membrane filter in an appropriate cassette
filter holder is used in conjunction with a personal sampling
pump capable of maintaining a face velocity of 2.6 cm/s. The
cassette is attached to the worker's lapel and connected to
the pump tubing. The pump is operated at 1.5 L/min. In general, a 2-h sample, equivalent to 180 L of air, will provide
sufficient material for detection at 20% of the threshold
limit values. The flow rate, ambient temperature, and
ambient pressure are recorded at the beginning and again at
the end of the sampling period.

At the end of the sampling period, the cassette is removed
from the worker's lapel and the end plugs are replaced. The
sealed cassette is retained for subsequent analysis of the
sample it contains.

The filter containing the sample is carefully removed from the cassette and transferred to a 125-mL beaker. Digestion of the samples and blanks is carried out in accord with the NIOSH procedure presented in Section 3.1.1.1.

The determinations of the metals are made using the information in Table 4.1 as a guide.

Collaborative testing of this method by 16 laboratories, using filters loaded at three concentration levels in a dynamic aerosol generation and sampling system, indicated that the average percentage recoveries and the standard deviations for the representative metals were: Cd, 100.8 ± 9.9; Co, 97.6 ± 13.9; Cr, 96.6 ± 10.8; Ni, 98.6 ± 10.3; and Pb, 98.7 ± 12.2. The relative standard deviation of the analytical measurements is approximately 3% when the measurements are made in the ranges listed in Table 4.2.

4.2. Procedure for Arsenic: NIOSH Method P&CAM 139

Procedure P&CAM 139 describes the collection, dissolution and determination of arsenic in airborne particulate matter. The sample is collected on a membrane filter and dissolved in nitric-sulfuric acid mixture, and arsenic is determined by hydride generation in the argon-hydrogen flame. For a 30-L air sample, the range of the method is from 0.002 mg/m^3 to 0.06 mg/m^3. The precision of the method shows a ±10.7% coefficient of variation.

The air sample is collected with a 37-mm, type AA, Millipore filter, or equivalent, contained in a suitable cassette. The filter assembly is connected to a personal sampling pump operated at 1.7 L/min. Ambient temperature and pressure are recorded at the beginning and again at the end of the sampling period. A minimum sample of 30 L should be collected.

The filter containing the sample is transferred to a 125-mL beaker and treated with 3 mL of concentrated nitric acid and 2 mL of concentrated sulfuric acid. The contents of the beaker are heated gently, and nitric acid is added as needed until a colorless liquid is obtained.

Continue heating until dense white fumes of SO_3 are produced. After cooling, transfer the colorless liquid to a 25-mL volumetric flask and make up to volume with distilled water.

A 5.00-mL aliquot of the sample solution is transferred to the hydride generator and treated with 25 mL of distilled water and 3 mL of concentrated hydrochloric acid. The hydride, arsine, is generated by the addition of sodium borohydride and flushed into the argon-hydrogen flame. Absorbance measurements are made at a wavelength of 193.7 nm with the aid of a strip chart recorder. The arsenic content of the sample is determined by reference to a calibration curve prepared by simultaneous measurements with standards.

4.3. Procedure for Beryllium: NIOSH Method P&CAM 288

Procedure P&CAM 288 describes the collection, dissolution, and determination of beryllium and beryllium compounds. Collection and dissolution are carried out as above, and the determination is made using the furnace technique. The range of the method is from 0.5 $\mu g/m^3$ to 10 $\mu g/m^3$ when a 90-L sample is collected. Recovery of beryllium was 98.2% and the coefficient of variation was 0.8% when National Bureau of Standards (NBS) standard reference material 2675, air filter, was tested.

Pass the air sample through a 37-mm diameter, 0.8-μm pore size mixed cellulose ester membrane filter secured in a suitable cassette. To determine whether the ceiling OSHA standard is exceeded, collect 25 L by sampling for 15 min at a flow rate of 1.7 L/min. Sample for 1 h at 1.7 L/min, equivalent to 90 L of air, to determine whether the time weighted average (TWA) OSHA standard is exceeded. Record the flow rate, temperature, and pressure at the beginning and the end of the sampling period.

Carefully transfer the filter and the sample it contains from the cassette to a 125-mL beaker. Treat the contents of the beaker with 10 mL of concentrated nitric acid and 1 mL of concentrated sulfuric acid. Cover with a watch glass and heat until the brown fumes from the nitric acid are no longer evolved. Continue heating until the dense white fumes of SO_3 appear. Cool and wash down the watch glass and inside of the beaker with distilled water. Evaporate the contents of the beaker just to dryness. Cool and add exactly 10 mL of 2% (w/v) sodium sulfate in 3% (v/v) sulfuric acid. Seal the beakers with parafilm, heat for 10 min in a 60°-70°C water

Table 4.1. Instrument Parameters

Element	Type of flame (o, oxidizing; r, reducing)	Analytical wavelength (nm)	Interferences[d]	Remedy[a]
Ag	Air-C_2H_2 (o)	328.1	IO_3^-, WO_4^{-2}, MnO_4^{-1}	b
Al[c]	N_2O-C_2H_2 (r)	309.3	Ionization, SO_4^{-2}, V, Fe, HCl, H_2SO_4	b, d, e
As	Air-C_2H_2 (o)	193.7	Background absorption	g
Ba	N_2O-C_2H_2 (r)	553.6	Ionization, large conc. of Ca	d, f
Be[c]	N_2O-C_2H_2 (r)	234.9	Al, Si, Mn	e, c, g
Bi	Air-C_2H_2 (o)	223.1		g
Ca	Air-C_2H_2 (r) N_2O-C_2H_2 (r)	422.7	Ionization and chemical	d, e
Cd	Air-C_2H_2 (o)	228.8		e, g
Co[c]	Air-C_2H_2 (o)	240.7		e, g
Cr[c]	Air-C_2H_2 (r)	357.9	Fe, Ni	e, b
Cu	Air-C_2H_2 (o)	324.8		e, b
Fe	Air-C_2H_2 (o)	248.3	High Ni conc., Si	b, g

Element	Flame		Wavelength	Interference	Notes
In	Air-C$_2$H$_2$	(o)	303.9	Al, Mg, Cu, Zn, H$_x$PO$_4^{x-3}$	b
K	Air-C$_2$H$_2$	(o)	766.5	Ionization	d
Li	Air-C$_2$H$_2$	(o)	670.8	Ionization	d
Mg	Air-C$_2$H$_2$	(o)	285.2	Chemical	e
	N$_2$O-C$_2$H$_2$	(o)		Ionization	d
Mn	Air-C$_2$H$_2$	(o)	279.5		
Mo	N$_2$O-C$_2$H$_2$	(r)	313.5	Ca and other ions	e, h
Na	Air-C$_2$H$_2$	(o)	589.6	Ionization	e
Ni	Air-C$_2$H$_2$	(o)	232.0		e, g
Pb	Air-C$_2$H$_2$	(o)	217.0 / 283.3	Ca, high conc. SO$_4^{-2}$	e, b, g
Pd	Air-C$_2$H$_2$	(o)	247.6	Al, Co, Ni, Pt, Rh, Ru	e
Rb	Air-C$_2$H$_2$	(o)	780.0	Ionization	d
Sb	Air-C$_2$H$_2$	(o)	217.6	Pb	i, g
Si	N$_2$O-C$_2$H$_2$	(r)	251.6	Avoided by not using multi-element lamp containing Fe	
Sr	Air-C$_2$H$_2$	(r)	460.7	Ionization and chemical	d, e
	N$_2$O-C$_2$H$_2$	(r)			
Te	Air-C$_2$H$_2$	(o)	241.3		g

Table 4.1. Continued

Element	Type of flame (o, oxidizing; r, reducing)	Analytical wavelength (nm)	Interferences[d]	Remedy[a]
Tl	Air-C_2H_2 (o)	276.8		
V	N_2O-C_2H_2 (r)	318.4		
Zn	Air-C_2H_2 (o)	213.9		g

[a] High concentrations of silicates in the sample can cause an interference for many of the elements in this table and may cause aspiration problems. No matter what elements are being measured, if large amounts of silicates are extracted from the samples, the samples should be allowed to stand for several hours and centrifuged or filtered to remove the silicates.

[b] Samples are periodically analyzed by the method of additions to check for chemical interferences. If interferences are encountered, determinations must be made by the standard additions method or, if the interferent is identified, it may be added to the standards.

[c] Some compounds of these elements will not be dissolved by the procedure described here. When determining these elements one should verify that the types of compounds suspected in the sample will dissolve using this procedure.

[d] Ionization interferences are controlled by bringing all solutions to 1000 g/mL Cs (samples and standards.

[e] 1000 g/mL solution of La as a releasing agent is added to all samples and standards.

[f] In the presence of very large Ca concentrations (greater than 0.1%) a molecular absorption from $Ca(OH)_2$ may be observed. This interference may be overcome by using background correction when analyzing for Ba.

[g] Use D_2 or H_2 continuum for background correction.

[h] Add 1000 g/mL Al to both standards and samples.

[i] Use alternate Sb line (231.2 nm).

Table 4.2. Sensitivity Data

Element	Sensitivity[a] (μg/mL)	Solution detection limits[a] (μg/mL)	Range[a,b] μg/mL	Range[a,b] μg/m³ [c,d]	Minimum threshold value (TLV) (μg/m³)[e,f]
Ag	0.06	0.002	0.5-4.0	21-170	10 (metal and soluble compounds
Al	1.0	0.02	5-50	210-2100	NL
As	0.8	0.2	10-50	420-2100	50 (arsenic trioxide production 500 (arsenic and compounds as As)
Ba	0.4	0.008	1-25	42-1050	500 (soluble compounds)
Be	0.025	0.001	0.1-2.0	4.2-84	2
Bi	0.5	0.025	1-30	42-1250	NL
Ca	0.08	0.005	0.1-5.0	4.2-210	2,000 (CaO)
Cd	0.025	0.001	0.1-2.0	4.2-84	50 (metal dust, soluble salts, cadmium oxide fume)
Co	0.15	0.01	0.5-5.0	21-210	50 (metal fume and dusts)

Element					
Cr	0.1	0.003	0.5–5.0	21–210	500 (soluble chromic, chromous salts) 100 (chromic acid and chromate as CrO_3) 100 (chromite ore processing, as CrO_3)
Cu	0.09	0.002	0.5–5.0	21–210	1,000 (dusts and mists) 200 (fume)
Fe	0.12	0.005	0.5–5.0	21–210	5,000 (iron oxide fume, as iron oxide)
In	0.7	0.02	5–50	210–2100	1,000 (soluble compounds)
K	0.04	0.002	0.1–2.0	4.2–84	100 (metal and compounds) 2,000 (as KOH)
Li	0.035	0.0003	0.1–2.0	4.2–84	25 (as lithium hydride)
Mg	0.007	0.0001	0.05–0.50	2.1–21	10,000 (as magnesium oxide fume)
Mn	0.055	0.002	0.5–3.0	21–125	5,000 (metal and compounds)
Mo	0.5	0.02	15–40	625–1650	5,000 (solubles, as Mo) 10,000 (insoluble, as Mo)
Na	0.015	0.0002	0.05–1.0	2.1–42	2,000 (as NaOH)

Table 4.2. Continued

Element	Sensitivity[a] (μg/mL)	Solution detection limits[a] (μg/mL)	Range[a,b] μg/mL	Range[a,b] $\mu g/m^3$ [c,d]	Minimum threshold value (TLV) ($\mu g/m^3$)[e,f]
Ni	0.15	0.005	0.5-5.0	21-210	100 (metal and soluble compounds)
Pb	0.5	0.01	1-20	42-840	150 (inorganic compounds, fumes and dusts)
Pd	0.25	0.02	4-15	170-625	NL
Rb	0.1	0.002	0.5-5.0	21-210	NL
Sb	0.5	0.04	10-40	420-1650	50 (antimony trioxide production)
					500 (antimony trioxide handling and use, as Sb)
Si	1.8	0.02	50-150	2100-6300	10,000
Sr	0.12	0.002	0.5-5.0	21-210	NL

Te	0.5	0.03	5-25	210-1050	100
Tl	0.5	0.01	5-20	210-840	100 (soluble compounds)
V	1.7	0.04	10-150	420-6300	500 (V_2O_5 dust) 50 (V_2O_5 fume)
Zn	0.018	0.001	0.1-1.0	4.2-42	5,000 (ZnO fume)

[a] Data from "Analytical Methods for Atomic Absorption Spectrometry," Perkin-Elmer Corp., Norwalk, Connecticut, 1976.

[b] Data from "Analytical Methods for Flame Spectroscopy," Varian Techtron, Australia, 1972.

[c] The atmospheric concentrations were calculated assuming a collection volume of 0.24 m^3 (2L/min. for 2 h) and an analyte volume of 10 mL for the entire sample.

[d] For elements whose TLV is less than the lower limits of the range, a larger sampling volume will be required.

[e] Threshold limit values of airborne contaminants and physical agents with intended changes adopted by American Council of Government, Industrial Hygienists (ACGIH) for 1976. All values listed are expressed as elemental concentrations except as noted.

[f] NL signifies no limit expressed for this element or its compounds.

bath, and allow the solutions to stand overnight before determining the beryllium content by furnace atomic absorption spectrometry.

A 10 μL aliquot of the sample solution or standard is injected into the graphite furnace. The drying, ashing, and atomizing conditions are: 110°C for 20 s, 900°C for 10 s, and 2800°C for 8 s. Absorbance is measured as a 3-s integrated peak at a wavelength of 234.9 nm and a spectral slit of 0.7 nm. Argon flow is interrupted for the 3-s integration only if pyrolytically coated furnace tubes are used. Deuterium arc background correction is necessary. Triplicate injections of the standards should be alternated with triplicate injections of the samples to establish the direct comparison relationship. The standards must be prepared in 2% sodium sulfate-3% sulfuric acid solution.

4.4. Procedure for Barium (Soluble Compounds)
NIOSH Method S198

Procedure S198 describes the collection, dissolution, and determination of water soluble barium compounds in airborne particulate matter. The sample is collected on a membrane filter and extracted from the filter with hot water, and its barium content is determined by atomic absorption spectrometry in the nitrous oxide-acetylene flame. This method has been validated over the range of 0.281 mg/m^3 to 1.084 mg/m^3 using a 168-L sample, and it is estimated to have a working range of from 0.15 mg/m^3 to 1.3 mg/m^3. The coefficient of variation for both sampling and analysis was 5.5%. The collection efficiency was 100%.

A mixed cellulose ester filter of 37-mm diameter and 0.8-μm pore size is used to collect the sample. This filter in a three-piece cassette holder is secured to the worker's lapel, the cassette plugs are removed, and connection to a personal sampling pump is made. A 180-L sample is collected at a sampling rate of 1.5 L/min. The flow rate should be checked frequently during the sampling period, and the ambient temperature and pressure should be recorded at its beginning and end.

Transfer the filter containing the sample to a clean 125-mL beaker. An unused filter should be similarly treated as a

blank. Add 10 mL of boiling water to each beaker and allow
to stand with occasional swirling for 10 min. Decant each
solution into a second separate beaker. Wash each original
beaker and filter twice with hot water, and add the washings
to the corresponding second beakers. Add another 10 mL of
boiling water to each original beaker and filter, allow to
stand with occasional swirling for 10 min, and decant and
wash as above. Remove each filter with forceps and wash each
side of the filter with a stream of hot distilled water into
the second beaker. Carefully rinse the original beakers three
times with hot distilled water into the corresponding second
beakers. Filter the contents of the second beakers to remove
insoluble barium compounds, add three drops of concentrated
hydrochloric acid to each filtrate, and evaporate each
filtrate to dryness. Cool the beakers and dissolve the residues in exactly 5 mL of dilute (5% v/v) hydrochloric acid
containing 1100 ppm sodium.

These solutions are aspirated into a reducing nitrous oxide-
acetylene flame, and the absorbances are measured at 553.6 nm.
The absorbances of standards prepared in 5% hydrochloric acid
containing 1100 ppm sodium are measured under identical conditions for the preparation of the calibration curve. The
barium content of the samples is determined from the calibration curve. It is advisable to measure an occasional standard
between every few samples to insure that conditions have not
changed.

4.5. Procedure for Cadmium (Dust):
NIOSH Method S312

The procedure described is for the collection, dissolution
and determination of cadmium and its compounds present as airborne dusts in the occupational environment. The sample is
collected on a membrane filter, nitric acid is used to destroy
the organic matrix, the cadmium is dissolved in hydrochloric
acid and determined by conventional atomic absorption spectrometry in the air-acetylene flame. The method was validated over
the range of from 0.12 mg/m^3 to 0.98 mg/m^3 using a 25-L sample.
The coefficient of variation for sampling and analysis in this
range was 6.4%. The efficiencies of collection and recovery
were 100% and 105%, respectively.

To collect cadmium dusts, a personal sampler pump is used to pull air through a 37-mm diameter, 0.8-μm pore size cellulose ester membrane filter. The pump is operated at a flow rate of 1.5 L/min, and ambient temperature and pressure are recorded at the beginning and end of the sampling period. The flow rates are also recorded at these times.

The filter and sample are transferred to a 125-mL beaker. An unused filter is placed in another 125-mL beaker and carried through the process as a blank. To each beaker are added 2 mL of concentrated nitric acid. The beakers are covered with watch glasses and heated on a hotplate in a fume hood until the filters are dissolved and the volume of acid is reduced to approximately 0.5 mL. Twice more, add 2 mL of nitric acid to the beakers and evaporate down to 0.5 mL. Add 2 mL of concentrated hydrochloric acid to each beaker and evaporate down to 0.5 mL. Repeat the hydrochloric acid addition and the evaporation to 0.5 mL twice more. Cool the solutions, add 10 mL of high-purity water, and quantitatively transfer to 25-mL volumetric flasks. Dilute the contents of each flask to volume.

Measure the absorbances of the samples and standards made up in 0.5 N hydrochloric acid at 228.8 nm in an oxidizing air-acetylene flame. The cadmium content of the samples is determined by direct comparison of their absorbances to those of the standards.

4.6. Procedure for Cadmium (Fume): NIOSH Method S313

Method S313 describes the collection and dissolution of samples for the determination of cadmium fume by atomic absorption spectrometry. The procedures are the same as those used for the collection, dissolution, and determination of cadmium and its compounds in airborne dusts, NIOSH method S312. Cadmium fume is differentiated from cadmium dust on the basis of particle size and source. An acceptable size cut off for fume is below 1 μm. Cadmium fume usually results from oxidation, sublimation, or distillation processes followed by condensation. Cadmium dust, on the other hand, is associated with mining and ore-reduction operations.

4.7. Procedure for Calcium (Oxide): NIOSH Method S205

Procedure S205 describes the collection, dissolution, and determination of calcium compounds in airborne particulates. The sample is collected by drawing a known volume of air through a membrane filter, the organic matrix is destroyed by digestion in nitric-perchloric acid mixture, and the calcium is dissolved in hydrochloric acid and determined by atomic absorption spectrometry in the air-acetylene flame. The method was validated over the range of 2.6 mg/m^3 to 10.16 mg/m^3 using an 85-L air sample. Under these conditions, the working range of the method is estimated to be from 1 to 20 mg/m^3. The collection efficiency was 100%, and the overall precision of the method was reflected in a coefficient of variation of 6.3%.

The sample is collected by passing air through a 37-mm, 0.8-μm pore size mixed ester membrane filter contained in a three-piece cassette. An 85-L sample is recommended, and the collection rate should be 1.5 L/min. The pump rotameter should be observed frequently during sample collection, and the ambient temperature and pressure should be recorded at the beginning and at the end of the sampling period.

Transfer the filter containing the sample to a 125-mL beaker. Blank filters should be carried through the process. Treat the contents of the beakers with 5 mL of concentrated nitric acid. Cover the beakers with watch glasses and heat until the filters are dissolved and most of the acid has evaporated. Cool and add 2 mL of concentrated nitric acid and 1 mL of 60% perchloric acid. Heat in a fume hood approved for perchloric acid until dense white fumes are evolved. Cool, wash down the watch glasses and the inside of the beakers, and evaporate the solutions to dryness. Dissolve the residues in 5 mL of dilute (5% v/v) hydrochloric acid containing 1% lanthanum. Transfer quantitatively to 100-mL volumetric flasks, and dilute to volume with 5% hydrochloric acid containing 1% lanthanum.

Aspirate the solutions into an oxidizing air-acetylene flame and measure the absorbances at 422.7 nm. Prepare a calibration curve by aspirating calcium standards in 5% hydrochloric acid containing 1% lanthanum under identical conditions and determine the calcium content of the samples by direct comparison.

4.8. Procedure for Cobalt (Dust and Fume): NIOSH Method S203

Procedure S203 describes the collection, dissolution, and determination of cobalt and cobalt compounds as dusts and fumes in the air of the occupational environment. The sample is collected on a mixed ester membrane filter, the sample matrix is destroyed, and the sample is dissolved by digestion in aqua regia; the cobalt content of the sample is then determined by conventional atomic absorption spectrometry in the air-acetylene flame. This method was independently validated for cobalt fume over the range of 0.031 mg/m^3 to 0.221 mg/m^3 and for cobalt dust over the range of 0.04 mg/m^3 to 0.262 mg/m^3. The working range of this method is estimated to be from 0.01 mg/m^3 to 0.3 mg/m^3. Collection efficiency of 100% and recovery of 102% were demonstrated. The coefficient of variation for collection, dissolution, and determination is 7%.

To collect cobalt metal, dust and/or fume, a personal sampling pump is used to pull the air sample through a 37-mm diameter, 0.8-μm pore size mixed ester membrane filter contained in a suitable cassette.

Sample at a flow rate of 1.5 L/min and collect a 3-h sample. Check the flow rate frequently during the sampling period and record the ambient temperature and pressure at the beginning and at the end of the sampling period.

Transfer the filter containing the sample to a 50-mL beaker. Similarly transfer a blank filter and filters spiked with known amounts of cobalt standard (0, 0.5, 1, and 2 times the OSHA standard) to separate 50-mL beakers. The former is used to determine cobalt impurities in the filters or reagents; the latter is used to determine recoveries of cobalt from the filters. Treat the contents of each beaker with 3 mL of aqua regia and allow to stand at room temperature for 30 min. Cover each beaker with a watch glass and heat the contents of each beaker until the filter is dissolved and the volume is reduced to 0.5 mL. Add 3 mL of concentrated nitric acid and heat to evaporate down to 0.5 mL. Add a second 3 mL of

nitric acid and continue heating until the volume is reduced
to 1 mL. Transfer the contents of the beakers quantitatively
to separate 10-mL volumetric flasks. Bring the contents of the
flasks to volume with 5% nitric acid.

Aspirate the solutions into an oxidizing air-acetylene flame
and record the absorbance at 240.7 nm. Prepare a calibration
curve by aspirating standards made up in 5% nitric acid under
identical conditions and determine the cobalt content of the
samples, blanks, and spiked filters by direct comparison.

4.9. Procedure for Copper (Dust and Mist):
NIOSH Method S186

Procedure S186 describes the collection, dissolution, and
determination of copper and copper compounds present as dusts
and mists in the work place. The samples are collected on
membrane filters, the matrix is destroyed with nitric acid
digestion, and the copper is dissolved in hydrochloric acid
and is determined by conventional atomic absorption spectrometry in the air-acetylene flame. The method has been validated
over the range of 0.468 mg/m^3 to 1.83 mg/m^3 using a 90-L
sample. Under these conditions, the working range of the
method is estimated to be from 0.1 mg/m^3 to 3 mg/m^3. The
collection efficiency and recovery were 99.6% and 101%, respectively, and the overall coefficient of variation for both
the sampling and the analysis was 5.1%.

Collect the sample on a 37-mm diameter, 0.8-μm pore size
mixed cellulose ester membrane filter contained in a suitable
cassette. Connect the filter cassette to a personal sampling
pump, and sample for 1 h at a flow rate of 1.5 L/min. Record
the ambient temperature and pressure at the start and finish
of the sampling period and check the flow rate frequently
during this time.

Carefully remove the filter from the cassette and transfer
it with the sample it contains to a 125-mL beaker. Similarly
transfer an unused filter and filters spiked with 0, 0.5, 1,
and 2 times the OSHA standard to separate 125-mL beakers.
Add 2 mL of concentrated nitric acid to the contents of each
beaker, cover the beakers with watch glasses, and heat until
the filters have dissolved and the volumes have been reduced
to 0.5 mL. Repeat the addition of nitric acid, heating, and

evaporation to 0.5 mL twice. Cool and add 2 mL of 6 N hydrochloric acid to each beaker. Heat to evaporate the hydrochloric acid down to 0.5 mL. Repeat the addition of hydrochloric acid, heating, and evaporation twice. Cool the solutions, add 10 mL of high-purity water and quantitatively
transfer to 25-mL volumetric flasks. Bring to volume with
high-purity water.

Prepare a calibration curve by aspirating copper standards
made up in 0.2 N hydrochloric acid. Aspirate the sample solutions, and the solutions from the blank and spiked filters.
All absorbance measurements are made at a wavelength of
324.7 nm. Determine the copper content of the samples, blanks,
and spikes by direct comparison.

4.10. Procedure for Copper (Fume): NIOSH Method S354

Procedure S354 describes the collection, dissolution, and
determination of copper fumes. Particulate matter containing
copper or copper compounds is collected simultaneously, and it
will result in high results being reported as copper fume.
This method has been validated over the range of 0.0548 mg/m^3
to 0.372 mg/m^3 using a 480-L sample. The working range of
the method is estimated to be 0.01 mg/m^3 to 0.3 mg/m^3. The
collection efficiency was determined to be 100%, and the coefficient of variation for the total analytical and sampling
method was 5.8%.

Collect the sample on a 37-mm diameter, 0.8-μm pore size
mixed cellulose ester membrane filter supported on a glass
fiber backup pad contained in an appropriate two-piece cassette
filter holder. Clip the cassette to the worker's lapel and
connect it to the sampling pump tubing. Sample at a flow rate
of 2 L/min for 4 h. Record the ambient temperature and pressure at the beginning and again at the end of the sampling
period and check the flow rate frequently during this period.

Carefully transfer each sample to separate 125-mL beakers,
treat with 10 mL of concentrated nitric acid, cover with watch
glasses, and heat until most of the acid has evaporated. Add
3 mL more of nitric acid and continue heating until the volumes are reduced to 0.5 mL. Cool, rinse down the watch glasses and insides of the beakers with 1% nitric acid, and

quantitatively transfer the solutions to 10-mL volumetric
flasks. Dilute the contents of the flasks with 1% nitric
acid.

Measure the absorbance of the samples and standards (also
made up in 1% nitric acid) at a wavelength of 324.7 nm by
aspiration into an oxidizing air-acetylene flame. The copper
content of the samples is determined by direct comparison.

4.11. Procedure for Chromium (Metal and Insoluble Compound): NIOSH Method S352

Procedure S352 describes the collection, dissolution, and
determination of chromium and chromium compounds in airborne
particulate matter from the work place. The method is limited
to those chemical forms of chromium soluble in hot, concentrated nitric acid. This method has been validated over the
range of 0.282 mg/m^3 to 0.947 mg/m^3 using a 90-L sample. On
this basis, the working range of the method is estimated to be
from 0.05 mg/m^3 to 2.5 mg/m^3. The collection efficiency was
100%, and the overall coefficient of variation was 8.5%.

The sample is collected on a 37-mm diameter, 0.8-μm pore
size mixed ester membrane filter contained in an appropriate
three-piece cassette filter holder and secured with a cellulose backup pad. Attach the filter cassette to the worker's
lapel, connect it to the personal sampling pump, and collect a
90-L sample at a flow rate of 1.5 L/min. Check the flow rate
and record the ambient temperature and pressure at the beginning and end of the sampling period.

Remove the filter from the cassette and transfer it to a
100-mL beaker. Add 2-3 mL of concentrated nitric acid to the
beaker, cover with a watch glass, and heat until the volume
of acid is reduced to approximately 0.5 mL. Repeat the addition of acid, heating, and evaporation process twice. Allow
the last evaporation to go to dryness and continue heating
until a white ash appears. Cool and wash down the watch
glass and inside of the beaker with distilled water. Again
evaporate to dryness. Cool and add 1 mL of concentrated
nitric acid. Quantitatively transfer the solution to a 20-mL
graduated centrifuge tube and bring to volume with high-
purity water.

Prepare a calibration curve by measuring the absorbance of

chromium standards in a reducing nitrous oxide-acetylene flame
at a wavelength of 357.9 nm. Aspirate the sample solutions
under identical conditions and determine their chromium content by direct comparison.

4.12. Procedure for Chromium (Soluble Compounds): NIOSH Method S323

Procedure S323 describes the collection, dissolution, and
determination of chromium and chromium compounds in airborne
particulate matter. This procedure parallels that cited in
Section 4.11 except that sample dissolution for the former is
achieved with both hydrochloric and nitric acids while dissolution of the latter makes use of only nitric acid. This method
was validated over the range of 0.493 mg/m^3 to 1.830 mg/m^3
using a 90-L sample, and the working range of 0.05 mg/m^3 to
2.5 mg/m^3 was estimated from these conditions. The collection
efficiency was 100%, and the overall coefficient of variation
for both collection and analysis was 7.6%.

Sample collection is the same as presented in Section 4.11.

Dissolution of the sample and destruction of the matrix is
as follows: treat the sample with 3 mL of 6 M hydrochloric
acid. Cover the beaker with a watch glass and heat until most
of the acid is evaporated. Repeat twice. Add 3 mL of concentrated nitric acid and continue as described in Section 4.11.

The determination of chromium is made as described in
Section 4.11.

4.13. Procedure for Chromium (Total Particulate): NIOSH Method P&CAM 152

Procedure P&CAM 152 describes the collection, dissolution,
and determination of total chromium in airborne particulates
in the occupational environment. The sample is collected on
a membrane filter, the matrix is destroyed and the chromium
solubilized by nitric acid digestion, and the chromium content of the sample is determined by atomic absorption spectrometry in the nitrous oxide-acetylene flame. Evaluation of the
procedure indicated good recoveries and a coefficient of
variation of less than 5% when 10 filters spiked near the
threshold limit were analyzed.

Collect the sample on a Millipore 37-mm HA-type filter secured in an appropriate two- or three-piece cassette filter holder. Connect the exit end of the filter holder to the pump and collect a minimum sample of 100 L at a flow rate of 2 L/min. Record ambient temperature and pressure at the start and finish of the sampling period and check the flow rate during the collection period.

Carefully transfer the filter from the cassette to a 125-mL beaker. Treat with 3 mL of redistilled nitric acid, cover with a watch glass, and heat to dryness. Continue adding nitric acid, 1 mL at a time, taking to dryness until no organic matter remains. Then add 0.5 mL of nitric acid and several milliliters of water, swirl, and place on a hotplate but do not take to dryness. Quantitatively transfer to a 10-mL graduated centrifuge tube and make up to volume with high-purity water.

Prepare a calibration curve by aspirating chromium standards made up in 5% nitric acid into the nitrous oxide-acetylene flame and measuring the absorbances at 357.9-nm wavelength setting and a spectral band pass of 0.3 nm. Measure the absorbances of the sample solutions under identical conditions and determine their chromium contents by direct comparison.

4.14. Procedure for Iron (Oxide Fume): NIOSH Method S366

Procedure S366 describes the collection, dissolution and determination of iron oxide fume in the work place. Iron and iron compounds other than the oxide that are collected simultaneously will lead to high results for iron oxide. The sample is collected on a membrane filter, the sample matrix is destroyed, and the iron compounds solubilized by digestion in hydrochloric and nitric acids, and the iron content is determined by conventional atomic absorption spectrometry. This method was validated over the range of 3.87 mg/m^3 to 18.19 mg/m^3 using a 145-L sample. The working range of this method is estimated to be from 1.3 mg/m^3 to 35.7 mg/m^3. The collection efficiency of the method was determined to be 100%, and the total coefficient of variation for the sampling and the analysis was 6.7%.

Collect the sample on a 37-mm diameter, 0.8-μm pore size

mixed cellulose ester membrane filter secured in a three-piece filter cassette holder with cellulose backup pad. Clip the cassette to the worker's lapel and connect the cassette to the personal sampling pump. Collect a 150-L sample at a flow rate of 1.5 L/min. Check the flow rate frequently and record the ambient temperature and pressure.

To separate 100-mL beakers, transfer the filters containing the samples, an unused filter to serve as a blank, and filters spiked with 0, 0.5, 1, and 2 times the OSHA standard to serve as positive controls for the evaluation of recovery efficiency. Add 3 mL of 1:1 hydrochloric acid to each beaker, cover with watch glasses, and heat on a hotplate until most of the acid has evaporated. Repeat the addition and evaporation of hydrochloric acid twice. Add 3 mL of concentrated nitric acid to each beaker and heat until most of the acid has evaporated. Repeat the addition and evaporation of nitric acid twice. Continue heating after the third evaporation until a white ash appears. Cool, and rinse down the watch glasses and insides of the beakers with distilled water. Again evaporate to dryness. Cool, and add 12.5 mL of concentrated nitric acid to each beaker. Quantitatively transfer the contents of each beaker to separate 250-mL volumetric flasks. Bring the contents of the flasks to volume with distilled water.

Prepare a calibration curve by aspirating iron standards made up in dilute nitric acid into an oxidizing air-acetylene flame and measuring the absorbances at 248.3 nm. Measure the absorbances of the samples, the blank, and the positive controls under identical conditions, and determine their iron content by direct comparison.

4.15. Procedure for Lead: NIOSH
Method P&CAM 214

Procedure P&CAM 214 describes the collection, dissolution, and determination of lead in airborne particulate matter. The sample is collected on a membrane filter, the matrix is destroyed and the lead solubilized by nitric acid digestion, and the lead content of the sample is determined by furnace atomic absorption spectrometry using the method of standard additions. The working range of the procedure is from

0.02 mg/m^3 to 0.4 mg/m^3, and the coefficient of variation for the analytical part of the method is 5.7%.

Collect the sample on a cellulose ester membrane filter equivalent to Millipore Type AA (0.8-μm pore size, 37-mm diameter) contained in a Millipore MAWP 037 AO, or equivalent, filter holder. A 10-min sample at a flow rate of 1.5 L/min is adequate. Record ambient temperature and pressure.

Remove the filters from the cassettes and transfer them to 125-mL beakers. Include an unused filter as a blank. Add 2 mL of concentrated nitric acid to each beaker, cover with watch glasses, and heat on a hotplate until nearly dry. Add 1 mL more of nitric acid and again take to near dryness. Quantitatively transfer to 15-mL centrifuge tubes with 1% nitric acid heating the contents of the beakers after each addition of the nitric acid. Cool the contents of the centrifuge tubes and dilute to the 5 mL mark with 1% nitric acid.

Remove three 1.0-mL aliquots from each centrifuge tube and transfer them to small test tubes identified as "A," "B," and "C." To tube "A," add 20 μL of lead standard containing 10 μg/mL and 20 μL of high-purity water. To tube "B," add 40 μL of lead standard containing 10 μg/mL. To tube "C," add 40 μL of high-purity water.

Measure the absorbance of all solutions at 283.3 nm by the furnace technique. Background correction is desirable. The measurements should be made in triplicate using 10-μL injections, and the absorbances should be recorded on a strip chart. The following drying, ashing, and atomizing conditions are recommended: 110°C for 20 s, 400°C for 15 s, and 2000°C for 10 s. The lead content of the sample solutions should be determined by the method of standard additions.

4.16. Procedure for Magnesium (Oxide Fume):
NIOSH Method S369

Procedure S369 describes the collection, dissolution, and determination of magnesium oxide fume in the occupational environment. Any magnesium compound collected will be measured as magnesium oxide and the results may be correspondingly high. The sample is collected on a membrane filter, the sample matrix is destroyed and the sample is solubilized by nitric

acid digestion, and the determination is made by conventional
flame atomic absorption spectrometry. This method was validated over the range of 7.48 mg/m^3 to 28.6 mg/m^3 using a
150-L sample. On this basis, the working range was estimated
to be from 1 mg/m^3 to 55 mg/m^3. The collection efficiency was
100%, and the total coefficient of variation for the sampling
and analysis was 6.2%.

The sample is collected on a 37-mm diameter, 0.8-μm pore
size mixed cellulose ester membrane filter backed by a cellulose pad in a three-piece filter cassette holder. The cassette is attached to the worker's lapel, and the connection
to the personal sampling pump is made. A 150-L sample collected at a flow rate of 1.5 L/min is recommended. The flow
rate is checked frequently during the collection period and
the ambient temperature and pressure are recorded at its beginning and end.

Transfer the filter containing the samples and a blank
filter to separate 100-mL beakers. Add 3 mL of concentrated
nitric acid to each beaker, cover with watch glasses, and
heat until most of the acid has evaporated. Repeat the addition of acid and evaporation steps twice. Continue heating
after the last evaporation until a white ash appears. Cool,
and wash down the watch glasses and insides of the beakers
with distilled water. Again evaporate to dryness. Cool. and
add 10 mL of concentrated nitric acid to each beaker. Quantitatively transfer the contents of each beaker to separate
100-mL volumetric flasks. Carry out the transfer and final
dilution with distilled water.

Aspirate the sample solutions into an oxidizing air-acetylene flame, and measure the absorbance at 202.5 (or 285.2) nm.
Measure the absorbances of magnesium standards in 10% nitric
acid under identical conditions for the preparation of a calibration curve. Determine the magnesium levels of the samples
by direct comparison.

4.17. Procedure for Manganese:
NIOSH Method S5

Procedure S5 describes the collection, dissolution, and
determination of manganese compounds in airborne particulates.
The sample is collected on a membrane filter, the matrix is

destroyed and the manganese solubilized by digestion with
nitric and hydrochloric acids, and the manganese content of
the sample is determined by conventional flame atomic absorption spectrometry. This method was validated over the range
of 2.5 mg/m^3 to 10 mg/m^3 using a 22.5-L sample. The working
range of the method was estimated to be in the range of 0.2
mg/m^3 to 20 mg/m^3 on this basis. The coefficient of variation for total analytical and sampling processes was 6.5%,
and the mean values showed a 9% negative bias.

Collect the sample on a 37-mm diameter, 0.8-μm pore size
mixed ester membrane filter secured in an appropriate three-
piece cassette filter holder. Sample at a flow rate of
1.5 L/min and collect several 15-min samples.

Transfer the filters containing the samples to separate
125-mL beakers. Include an unused filter as a blank. Add
2 mL of concentrated nitric acid to each beaker, cover with
watch glasses, and heat gently until the filters are dissolved and most of the acid has evaporated. Repeat the addition of acid and evaporation to 0.5 mL twice. Complete the
digestion by three successive additions of 2 mL of 1:1 hydrochloric acid followed by evaporation to 0.5 mL after each
addition. Cool the solutions and add 10 mL of distilled
water to each. Transfer to 100-mL volumetric flasks quantitatively with distilled water and bring to volume with distilled water.

Aspirate the solutions into an oxidizing air-acetylene
flame and record the absorbances at 279.5 nm. Prepare a calibration curve by measuring the absorbances of manganese
standards in 0.3 N hydrochloric acid under identical conditions. Determine the manganese content of the samples by
direct comparison.

4.18. Procedure for Mercury: NIOSH
Method P&CAM 175

Procedure P&CAM 175 describes the selective collection and
determination of mercury in airborne particulates, as gaseous
organomercury compounds, and as mercury vapor. A special
sampling train (Figure 4.1, see page 106) consisting of a
glass fiber filter and a two-stage sampling tube is used to
selectively collect the sample. Particulate matter is

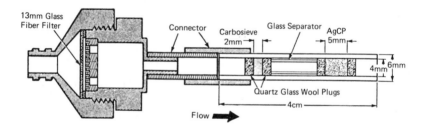

Figure 4.1. Two-stage mercury sampling
tube and glass fiber filter.[10]

retained on the filter. The first stage of the sampling tube
contains Carbosieve B* which selectively absorbs the gaseous
organomercury compounds. The second stage of the sampling
tube contains silvered Chromosorb P (AgCP) to trap the mercury
vapor. After the sample is collected, the sampling train is
separated into its three sections. The mercury in each individual section is converted to mercury vapor by thermal
decomposition or desorption and flushed into the absorption
cell of an atomic absorption spectrometer set up for the cold
vapor technique. Special details on the construction of the
thermal desorption apparatus are appended to this procedure.

Using a 3-L sample and a 6-in. absorption cell, the working range of the method is from 0.5 $\mu g/m^3$ to 0.5 mg/m^3. The
overall sampling and analytical precision for the various
forms of mercury is presented in Table 4.3. The pooled precision for the preparation and analysis of four groups of 10
tubes each, spiked at 20, 41, 81, and 122 ng of Hg per tube
and measured over a 2-week period, was reflected in a coefficient of variation equal to 5.6%.

Prepare the sampling train with a 13-mm diameter, 0.3-μm
pore size glass fiber filter in a filter holder cassette followed by a two-stage sorbent tube containing the Carbosieve B
in the forward section and the AgCP in the rear section.†

---

*Carbosieve B is available from Supelco, Inc., Bellefonte, PA.
  Prefilled tubes are available from SKC, Inc., Eighty Four, PA.

†The following items are available from SKC, Inc., RD 1, 395
  Valley View Road, Eighty Four, PA 15330: Cat. no. 225-15,
  13-mm, 0.3-μm, type AE glass fiber filter; Cat. no. 225-32,
  13-mm filter cassette; Cat. no. 226-17-2, two-stage sorbent
  tube.

Table 4.3. Precision of Mercury Measurements

Chemical form	Number of samples	Airborne concentration (mg/m^3)	Coefficient of variation (%)
Phenyl mercuric acetate (particulate)	6	0.110	10.7
Dimethyl mercury (organomercury)	9	0.058	7.4
Mercury vapor	13	0.065	7.3

Cycle the tube through the desorption unit to purge impurities before use and store it with the ends capped. Collect a 3-L sample at a flow rate of 100 mL/min. Record the ambient temperature and pressure at the time of sampling. At the end of the sampling period, immediately connect an unused sampling train to the inlet end of the original sampling train. Using the unused sampling train as a filter, purge the original sampling train for 10 min at a flow rate of 5 L/min.

The purge step is necessary to insure passage of mercury vapor through the filter and Carbosieve B on to the AgCP of the original filter train. Under these conditions, 5% of the mercury vapor is retained by the Carbosieve B. Hence, the quantity of mercury found in the AgCP section must be increased by a correction factor of 1.05. Correspondingly, the quantity of organomercury compounds must be reduced by an equal amount to the difference between the corrected mercury vapor value and the amount of mercury vapor actually found.

Discard the sampling train used as a filter and cap the original sampling train. Radiotracer experiments indicate that there are no losses of mercury from the sampling train during 7 days of storage at a temperature of 50°C. Nonetheless, the original sample tube should be analyzed for mercury as soon as possible.

Each of the three sections of the sampling train is analyzed for mercury separately. To avoid possible contamination, samples should not be touched with the bare hand. The analyses are performed by the cold vapor technique using either an atomic absorption spectrometer or a dedicated instrument such as the Coleman Mercury Analysis System (MAS-50).

Remove the glass fiber filter from its holder, fold it twice, and slide it into an unused, two-stage sampling tube from which the Carbosieve B, but not the AgCP, has been removed. Replace the glass wool plug behind the filter. Unclamp the loading mechanism of the decomposition unit, insert the tube containing the filter, push it up into the first desorption section and secure the loading mechanism. Proceed with the measurement of particulate mercury as outlined below.

Score and break the two-stage sampling tube between the second and third glass wool plugs. Unclamp the loading mechanism, insert the Carbosieve B stage, push it up into the first desorption section, and secure the loading mechanism. Proceed with the measurement of organic mercury as outlined below.

Unclamp the loading mechanism, insert the AgCP stage, push it up into the first desorption section, and secure the loading mechanism. Proceed with the measurement of mercury vapor as outlined below.

Prepare six calibration standards by pipetting 0, 20, 40, 60, 80, and 100 µL of standard solution containing 1 µg/mL of mercury directly on to the exposed AgCP of unused two-stage sample tubes from which the Carbosieve B has been removed. Replace the glass wool plugs, dry the tubes in an upright position for 6 h at 50°C, and measure the absorbances as outlined below.

Turn on the MAS-50 and allow it to stabilize.

Turn on the heating tape.

After the MAS-50 has stabilized, open the valve to obtain air flow through the system. Observe the rotameter and adjust the valve for the air flow to 1 L/min. Turn on the cooling air and adjust to 15 L/min.

Turn on the recorder and allow to stabilize. Adjust the MAS-50 and recorder to the desired 0% and 100% transmission settings.

Purge the desorption apparatus before measuring the absorbance of any sample or standard. Heat the first desorp-

tion section for 45 s and then heat the second desorption
section for 20 s.

Allow the desorption unit to cool for 7 min before insert-
ing a sample or standard. After this time, insert the sample or standard, push it up into the first desorption section,
and secure the loading mechanism.

Place an unused AgCP tube in the recollecting device. In
the event a sample reads off scale, it can be recovered in
the AgCP tube and measured again using a shorter path length
absorption cell in the optical path.

Turn on the first section of the thermal desorption unit,
and heat for 45 s. Wait 1 min, switch the three way power
switch to the second desorption section of the unit, and heat
for 25 s. Turn the three way switch to the off position.

Allow the thermal desorption unit to cool for 30 s, open
the loading mechanism, and allow the hot sample or standard
tube to drop into a dry beaker.

Allow the thermal desorption unit to cool for at least
7 min and reload with the next sample or standard.

Replace the AgCP tube in the recollection device and repeat
the heating cycle.

Prepare a calibration curve from the absorbance readings
obtained with the standards. Determine the particulate, organic, and elemental mercury levels in the samples by direct
comparison.

4.18.1. Appendix: Description and Installation of Two-Stage Thermal Desorption Unit

With the exception of the electrical components and the
loading spring, the entire thermal desorption unit is made of
either quartz or Pyrex glass. A diagram of the unit giving
the critical dimensions is shown in Figure 4.2, and details
on the loading mechanism are shown in Figure 4.3. The entire mercury analysis system is presented in Figure 4.4.
Each important part of the thermal desorption unit shown in
Figure 4.2 is numbered, and the numbers represent the
following:

   1 and 2. Loading mechanism (Figure 4.3). The loading
            mechanism is made from an 18/7 female glass joint
            with a steel spring and a plunger cut to reach
            the sample desorption section.

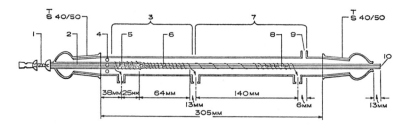

1. LOADING MECHANISM
2. QUARTZ GLASS PLUNGER TUBE
3. SAMPLE DESORPTION SECTION
4. COOLING AIR VENT HOLES
5. SAMPLING TUBE

6. CUPRIC OXIDE (CuO)
7. SECOND DESORPTION SECTION
8. GOLD SECTION
9. COOLING AIR INTAKE
10. OUTLET TO SPECTROPHOTOMETER

Figure 4.2. Thermal desorption unit.

3. Sample desorption section. The sample desorption
   section is made from 8-mm i.d. and 5-mm i.d. quartz
   tubing. The junction between the two sizes of
   tubing is tapered on the inside so that sampling
   tube tips fit snugly against the junction.
   Twenty-eight coils of 18-gage nichrome wire are
   wrapped around the first desorption section to
   heat the sampling tubes during the thermal desorption step. The nichrome wire must be evenly
   wrapped with approximately 5-mm spacing between
   coils.

4. Cooling vent holes. About seven 4-mm holes are
   placed around the outer jacket of the unit to
   allow cooling air to flow from the second heating
   coil to the first coil.

5. Sampling tube. Sampling tubes are positioned inside the first desorption section during mercury
   desorption. The spring on the loading mechanism
   is adjusted to press the sampling tubes lightly
   in place.

6. Cupric oxide (CuO). A 40-mm section of rod-shaped
   cupric oxide is placed just down stream from the
   sampling tube. The cupric oxide is held by a
   quartz glass wool plug which fits against a crimp
   in the glass tubing. Although the plugs should
   be large enough to hold the materials, they should
   not be packed too tightly.

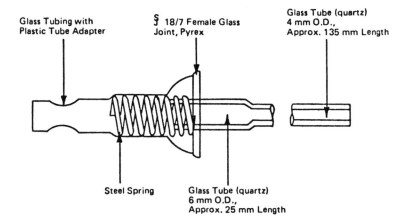

Figure 4.3. Loading mechanism.

7. Second desorption sections. This section is made
from the 5-mm i.d. quartz glass tubing extending
from the first desorption section. The contents
are held in place by a crimp in the glass tube
and quartz glass wool plugs. Thirty coils of size
20 nichrome wire are wrapped around the second
desorption section. The coils are wrapped to
allow most of the generated heat to concentrate
over the gold granules of the second desorption
section; ie, evenly spaced with 1-mm gaps between
the coils.

8. Gold section. This section consists of a 25-mm
length of 35/50 mesh granular gold mixed one to
one with 20/40 mesh sea sand. The sand is added
to the gold to prevent fusing of the gold granules
and to allow better air flow through the section.
This section must be at least 15 mm below the en-
trance of the cooling air.

9. Cooling air intake. A 6-mm i.d. piece of glass
tubing is used for connecting plastic tubing from
the cooling air supply to the cooling jacket.
The Pyrex glass cooling jacket not only directs
the flow of cooling air, but it also acts as electrical insulation for the heating coils. Wires
to the heating coils enter the cooling jacket

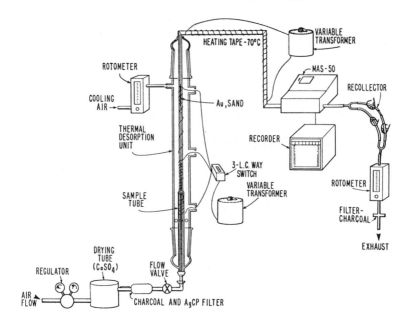

Figure 4.4. Mercury analysis system.

through 5-mm holes at the ends of glass nipples on the side of the jacket. The solderless connectors between the wires and the coils are placed inside the jacket to prevent exposing uninsulated wire outside the cooling jacket. The insulated wires to the connectors are sealed in place with a heat resistant sealer.

10. Outlet. The outlet from the desorption unit is butt connected with a Tygon overseal to a glass tube which leads to the cell of the atomic absorption spectrometer. The end of the quartz tube slides through an opening at the end of the cooling jacket. The opening is kept to a minimum to limit the escape of cooling air.

A diagram of the installed thermal desorption unit system is shown in Figure 4.4. The essential features of this diagram are the power connections, the air supply, the cooling air, and the detection system.

The wires to the heating coils should be heavy enough to
carry the current necessary to heat the coils. Power to the
heating coils is controlled by a three-position switch with
an off position, an on position for the heating coil at the
first desorption section and an on position for the heating
coil at the second desorption section. Power to this switch
comes from a 120-V ac input, 22-a variable transformer set at
28 V ac. The voltage should be set so that desorption temperatures of 500° ± 25°C is achieved in the sample tube.

The air supply which passes through the heated sections of
the thermal desorption unit is supplied at 8 psi. Before
entering the desorption unit, the air passes through a drying
tube containing anhydrous calcium sulfate. The dimensions of
the drying tube and all subsequent filters may vary, but they
should not interfere with the air flow through the system.
The calcium sulfate should be changed periodically depending
on the humidity of the air. The dried air passes through a
filter containing activated charcoal and 30/60 mesh silvered
Chromosorb P to remove mercury. The air then passes through
a needle valve, into the desorption unit, through the optical
cell of the atomic absorption spectrometer, through the rotameter, and into the final charcoal filter.

The cooling air flow is controlled at 15 L/min. The cooling air enters the desorption unit from a plastic tube into
the air intake near the second desorption stage.

The detection system is an atomic absorption spectrometer.
The mercury is desorbed from the gold (second) section of the
thermal desorption unit and enters the optical cell of the
spectrometer through a glass tube. This glass tube is maintained at 70°C by means of a heat tape. Power for the heat
tape is supplied by a 120-V ac, 8-a variable transformer.

4.19. Procedure for Mercury (Organic): NIOSH Method S342

Procedure S342 describes the collection and determination
of gaseous organomercury compounds in the air of the occupational environment. The sample is collected in an absorption
tube containing Carbosieve B, the organomercury compound is
desorbed and thermally decomposed in the thermal desorption
unit described in Section 4.18, and the resulting mercury

vapor is flushed into the optical cell of a cold vapor atomic absorption spectrometer. This method was validated over the range of 0.02 mg/m^3 to 0.08 mg/m^3 using a 3-L sample, and over the range of 0.004 mg/m^3 to 0.017 mg/m^3 using a 12-L sample. The coefficient of variation for the total analytical and sampling method for the smaller sample was 10%; for the larger sample, it was 7%. This method was found to be capable of yielding results within ±25% of the references values 95% of the time during the validation. The results showed a negative bias of 6.8% relative to the dynamically generated test concentration of dimethyl mercury. Samples collected from a test atmosphere containing 0.0107 mg/m^3 and stored for 7 days gave a mean value within 6.3% of that obtained from samples of the same atmosphere analyzed immediately after collection. Hence, the samples appear to be stable for at least 7 days after collection.

The sample is collected in a 2-cm-long, 6-mm o.d., 4-mm i.d. tube containing 12 mg of 45/60 Carbosieve B. A plug of quartz wool is placed at each end of the tube, and the tube is run through a desorption cycle before use. If other forms of mercury are expected to be present, the Carbosieve B tube should be protected with AgCP sorbent tube for elemental mercury and/or a 0.8-μm, 37-mm diameter, mixed cellulose ester membrane filter for particulate mercury. The Carbosieve B tube should be placed nearest the sampling pump, and either a 3-Liter sample or a 12-Liter sample should be collected at a flow rate between 0.01 L/min and 0.2 L/min. The temperature, pressure, and relative humidity of the atmosphere being sampled should be recorded, and the flow rate should be accurately known. After sampling, the AgCP tube and the membrane filter should be discarded and the Carbosieve B tube should be returned to its storage vial and retained for the analysis of organomercury levels.

The desorption apparatus should be purged before measuring the absorbance of any sample or standard. The sequence of operation for the desorption unit parallels that of Section 4.18 except that the heating times for the first and second sections are 60 s and 30 s, respectively, and the cooling time between cycles is 5 min. The absorbance signals are recorded as sharp peaks on the strip chart. Organomercury levels are determined by direct comparison. The standards are prepared by pipetting aliquots of standard solution on to the Carbo-

sieve B in unused sample tubes. Six standards are used: 0-,
2-, 4-, 6-, 8-, and 10-μL aliquots of 0.03-mg/mL mercury standard solution in 1% nitric acid.

4.20. Procedure for Molybdenum (Soluble Compounds): NIOSH Method S193

Procedure S193 describes the collection, dissolution, and
determination of soluble molybdenum compounds in the airborne
particulates of the work place. The sample is collected on a
membrane filter, soluble molybdenum compounds are dissolved in
water, variable enhancement effects by aluminum are minimized
by matrix matching, and the molybdenum content of the solutions
is measured by atomic absorption spectrometry in the nitrous
oxide-acetylene flame. This method was validated over the
range of 1.96 mg/m^3 to 9.75 mg/m^3. The sample size was 90-L,
and the working range of the method was estimated to be from
0.5 mg/m^3 to 15 mg/m^3. The coefficient of variation for the
total analytical and sampling method was 8.9%. The collection
efficiency was determined as 100%, and the average recovery
was found to be 101%.

Use a 0.8-μm pore size, 37-mm diameter mixed cellulose pad
contained in a three-piece cassette filter holder to collect
the sample. Sample at a flow rate of 1.5 L/min for 1 h to
collect a 90-L sample. Record the ambient temperature and
pressure during the sampling period.

Transfer the filters containing the samples, blank filters,
and filters spiked with 0.5, 1, and 2 times the OSHA standard
to separate 125-mL beakers. Add 10 mL of hot (boiling) distilled water to each beaker and allow to stand with occasional
swirling for 10 min. Quantitatively transfer the water extracts, using at least two 5-mL portions of hot water, to a
25-mL volumetric flask. Add 2.5 mL of 1 N nitric acid and
1 mL of a solution containing 10,000 μg/mL of aluminum to each
flask, and bring to volume with distilled water. Measure the
absorbances of samples and standards made up in 0.1 N nitric
acid containing 400 ppm aluminum in the nitrous oxide-acetylene
flame at 313.3 nm. Determine the molybdenum content of the
samples, the recoveries of the spiked filters, and the blank
values by direct comparison.

4.21. Procedure for Molybdenum (Insoluble Compounds): NIOSH Method S376

This procedure describes the collection, dissolution, and determination of molybdenum in airborne particulates from the occupational environment. The sample is collected on a membrane filter, decomposed by either nitric acid-perchloric acid digestion or alkaline hydrolysis followed by solubilization in nitric acid, and determined by atomic absorption spectrometry in the nitrous oxide-acetylene flame. This method was validated over the range of 6.3 mg/m^3 to 30.6 mg/m^3 using a 90-L sample. The working range of the method is estimated to be from 2 mg/m^3 to 50 mg/m^3. The coefficient of variation for the total analytical and sampling method was 5.6% when the dissolution was by acid digestion and 4.9% when alkaline hydrolysis was used to dissolve the sample. Collection efficiency was 100%, and recoveries were 100% for acid digestion and 98.5% for alkaline hydrolysis.

Collect the sample on a 0.8-μm pore size, 37-mm diameter mixed cellulose ester membrane filter secured in a three-piece cassette filter holder with a cellulose backup pad. Sample at a rate of 1.5 L/min for 60-min. Record the ambient temperature and pressure during the sampling period.

Collect small portions of the molybdenum compounds used in the work place and test their solubilities in acid and in alkali to determine which dissolution process is most applicable to the sample.

Transfer the filters with the samples, blank filters, and filters spiked with molybdenum at 0.5, 1, and 2 times the OSHA standard to the 125-mL beakers.

Use the following digestion for acid soluble molybdenum compounds (preferred whenever feasible). Add 2 mL of concentrated nitric acid to the contents of each beaker, cover with watch glass, and heat until the filters dissolve and the volumes are reduced to 0.5 mL. Add an additional 2 mL of nitric acid and again heat and evaporate to 0.5 mL. Cool and add 2 mL of concentrated nitric acid and 1 mL of concentrated perchloric acid to the contents of each beaker. Continue evaporation to fumes to effect complete solution. Do not allow the solutions to go to dryness during the evaporation. Add 5 drops (0.2 mL) of concentrated sulfuric acid to the contents of each beaker and allow to cool.

Use the following alkaline hydrolysis for base soluble molybdenum compounds (with caution and only when necessary). Add 5 mL of 2 N sodium hydroxide solution to the contents of each beaker and heat gently with occasional swirling. When the samples appear to have dissolved, cool the solutions on an ice bath and slowly add 5 mL of concentrated nitric acid. Heat cautiously with occasional swirling to white fumes. Add 3 mL more of nitric acid and heat to complete the digestions. White crystals of sodium nitrate which will dissolve at a later stage may appear at this point. Allow the solutions to cool and add 5 drops of concentrated sulfuric acid to each.

Dilute the contents of the beakers (from either of the above solubilization steps) with 10 mL of distilled water and quantitatively transfer to 50-mL volumetric flasks. Add 2 mL of 10,000-ppm aluminum solution to each flask and dilute to 50 mL with distilled water. Aspirate the solutions from the samples, spiked filters, blank filters, and standards containing 400 ppm aluminum into the nitrous oxide-acetylene flame and measure the absorbances at 313.3 nm. Use the direct comparison to determine the recoveries, blanks, and molybdenum levels of the samples.

4.22. Procedure for Nickel (Particulate and Compounds): NIOSH Method P&CAM 298

This procedure describes the collection, dissolution, and determination of nickel in the airborne particulates of the occupational environment. The sample is collected on a membrane filter, the matrix is destroyed and the nickel solubilized by digestion in nitric and perchloric acids (hydrofluoric acid is added if it is necessary to destroy silicates), and nickel is determined by furnace atomic absorption spectrometry. The working range of the method, based on a 200-L sample, is from 0.005 mg/m^3 to 0.030 mg/m^3. The precision of the analytical method, as determined by the analysis of three standards in triplicate, was 10.3%. Interlaboratory precision for two replicate field samples analyzed by three different laboratories ranged from 6.7% to 26%. Recovery of nickel from spiked filters was 90.4% ± 4.0% in one laboratory and 101% ± 3.5% in another.

Collect the sample on a 0.8-µm pore size, 37-mm diameter

mixed cellulose ester membrane filter with a cellulose backup pad contained in a suitable plastic cassette filter holder. Remove the plugs from the cassette and attach it to the personal sampling pump by means of flexible tubing. Clip the cassette, face down, to the worker's lapel. Sample at a rate of 1.5 L/min for a sufficient period of time to correspond to 200 to 400 L. Check the flow rate frequently during this sampling period. Record the sampling time, flow rate, and ambient temperature and pressure.

Open the cassette filter holders, carefully transfer the filters to 125-mL beakers, and discard the cellulose backup pads. Add 5 mL of concentrated nitric acid to each beaker, cover with watch glasses, and heat on a low-temperature hotplate for 6 to 12 h. Allow the acid to evaporate to a few drops; add 2 mL of nitric acid, 1 mL of perchloric acid, and 5 drops of 50% hydrofluoric acid to each beaker; cover with watch glasses; and heat until dense white fumes are evolved. Wash down the watch glasses and insides of the beakers, and evaporate to dryness. Cool, and dissolve, the residues in a few mL of 2% nitric acid. Quantitatively transfer the solutions to 25-mL volumetric flasks with 2% nitric acid and dilute to volume with 2% nitric acid.

Measure the absorbances of triplicate injections of the samples and standards at 341.5 nm using background correction. Depending on atomizer dimensions, aliquots of from 5 μL to 50 μL may be injected. Absorbance measurements may also be made at 232.0 nm, but background absorbance is higher at the shorter wavelength. Background correction can be made with either the deuterium lamp or with the nonabsorbing 231.5 nickel line if the 232.0 resonance line is used for the absorbance measurements. Midrange standards should be remeasured between every few samples. The following time-temperature program is recommended: dry, 20 s at 100°C; ash, 20 s at 700°C; atomize, 7 s at 2800°C. The nickel content of the samples is determined by direct comparison. If the matrix has a high dissolved solids, the method of standard additions is recommended.

4.23. Procedure for Rhodium (Metal, Dust, and Fume): NIOSH Method S188

Procedure S188 describes the collection, dissolution, and

determination of rhodium in the airborne particulates of the work place. The sample is collected on a membrane filter, the matrix is destroyed and the rhodium solubilized by digestion in nitric and hydrochloric acids, and the determination of rhodium is made by conventional atomic absorption spectrometry in the air-acetylene flame. This method was validated over the range of 0.057 mg/m^3 to 0.21 mg/m^3 using a 720-L sample. Under these conditions, the working range of the method is estimated as from 0.005 mg/m^3 to 0.21 mg/m^3. The coefficient of variation for the total analytical and sampling method was 7.9%. The collection efficiency and recovery are reported to be 100% and 94.7%, respectively.

Collect a 720-L sample at a flow rate of 1.5 L/min on a 0.8-μm pore size, 37-mm diameter mixed cellulose ester membrane filter contained in an appropriate three-piece cassette filter holder. Check the flow rate through the filter frequently during the 8-h sampling period. Record the ambient temperature and pressure at hours 0, 4, and 8 of the sampling period.

Carefully transfer the filters containing the samples and unused filters spiked with 0, 0.5, 1, and 2 times the OSHA standard to 125-mL beakers. Treat the contents of each beaker with 2 mL of concentrated nitric acid, cover with watch glasses, and heat on a low-temperature hotplate until the filters are dissolved and the volumes are reduced to 0.5 mL. Repeat the addition of acid and heating twice. Add 2 mL of 6 N hydrochloric acid and evaporate to 0.5 mL. Repeat this addition and evaporation twice. Cool the solutions and add 10 mL of high-purity water to each. Quantitatively transfer the solutions to 25-mL volumetric flasks. Add 2.5 mL of 30% (w/v) potassium bisulfate solution to each flask and dilute the contents to 25 mL with high-purity water. Aspirate the solutions from the samples, the spiked filters, and the standards, also made up in 0.2 N hydrochloric acid and 3% potassium bisulfate, into an oxidizing air-acetylene flame and measure the absorbances at 343.5 nm. Prepare a calibration curve from the absorbances of the standards and determine the rhodium contents of the samples by direct comparison.

4.24. Procedure for Selenium: NIOSH Method S190

Procedure S190 describes the collection, dissolution, and

determination of selenium in the airborne particulate matter
of the occupational environment. The sample is collected on a
membrane filter, the selenium is extracted from the filter
with dilute nitric acid, and the determination of selenium is
made by atomic absorption spectrometry in the argon-hydrogen
flame. This method was validated over the range of 0.10 mg/m^3
to 0.5 mg/m^3 using a 360-L sample. On this basis, the working
range of the method was estimated to be from 0.05 mg/m^3 to
0.60 mg/m^3. The coefficient of variation for the total analysis and sampling was 9%. The collection efficiency was 100%
and the recovery was 92%. No bias is introduced if corrections for the recovery are applied.

Collect a 360-L sample on a 0.8-μm pore size, 37-mm diameter mixed cellulose ester membrane filter at a flow rate of
1.5 L/min. Record ambient temperature and pressure at the beginning and end of the sampling period and check the flow rate
frequently during this time.

Carefully transfer the filters containing the samples and
filters spiked at 0, 0.5, 1, and 2 times the OSHA standard
to 125-mL beakers. Treat the contents of each beaker with
10 mL of 0.1 N nitric acid and allow to stand for 20 min with
occasional swirling. Quantitatively transfer the supernatant
liquid to 25-mL volumetric flasks. Repeat the extractions of
the filters using two 5-mL portions of 0.1 N nitric acid, and
quantitatively transfer these extracts to the appropriate
volumetric flasks. Dilute the contents of the flasks to 25 mL
with 0.1 N nitric acid.

Prepare the atomic absorption spectrometer with an electrodeless discharge lamp for selenium and the argon-hydrogen-
air flame. Aspirate the solutions from the extractions of the
sample filters, those of the spiked filters, and the selenium
standards made up in 0.1 N nitric acid. Measure the absorbances at a wavelength of 196.0 nm. Prepare a calibration
curve from the absorbances of the standards and determine the
selenium contents of the samples and the recovery of selenium
from the spiked filters by direct comparison.

4.25. Procedure for Tellurium:
NIOSH Method S204

Procedure S204 describes the collection, dissolution, and

determination of tellurium in the airborne particulate of the
work place. The sample is collected on a membrane filter,
the sample matrix is destroyed and the tellurium solubilized
by digestion in nitric and perchloric acids, and the tellurium
content of the sample is determined by atomic absorption
spectrometry in the air-acetylene flame. This method was
validated over the range of 0.05 mg/m^3 to 0.24 mg/m^3 using a
670-L sample. The working range of the method was estimated
as from 0.02 mg/m^3 to 0.4 mg/m^3. The coefficient of variation for the total analytical and sampling method was 5.5%,
and the collection efficiency was 100%.

Collect the sample on a 37-mm diameter, 0.8-μm pore size
mixed cellulose ester membrane filter. Secure the filter in
a three-piece filter cassette holder with a cellulose backup
pad. Sample at a flow rate of 1.5 L/min. A 670-L sample is
recommended. Check the flow rate frequently during the sampling period. Record the ambient temperature and pressure at
0, 4, and 7.5 h.

Transfer the filters containing the samples to separate
125-mL beakers and treat each with 5 mL of concentrated
nitric acid. Cover each beaker with a watch glass and heat
on a low-temperature hotplate until most of the acid has
evaporated. Add 2 mL of nitric acid and 1 mL of perchloric
acid to each beaker, cover, and heat until dense white fumes
of perchloric acid are evolved. Wash down the watch glasses
and the insides of the beakers with distilled water, and
allow the solutions to evaporate to dryness. Add exactly 5 mL
of dilute nitric acid to each beaker.

Prepare a calibration curve by measuring the absorbances
of tellurium standards in dilute nitric acid at 214.3 nm in
an oxidizing air-acetylene flame. Aspirate the solutions
from the filters under identical conditions and measure the
absorbances at 214.3 nm. Determine the tellurium contents of
the samples by direct comparison.

4.26. Procedure for Tellurium Hexafluoride:
NIOSH Method S187

Procedure S187 describes the collection, dissolution, and
determination of tellurium hexafluoride vapor in the work
place. The sample is collected on a charcoal filter, desorbed

into dilute sodium hydroxide, and determined by conventional atomic absorption spectrometry in the air-acetylene flame. This method was validated over the range of 0.094 mg/m^3 to 0.373 mg/m^3 using a 390-L sample. The working range of the method was estimated to be from 0.04 mg/m^3 to 0.6 mg/m^3. The coefficient of variation for the total analytical and sampling method was 5.4%, and the collection efficiency was 100%.

The sample is collected on activated charcoal prepared from coconut shells fired at 600°C. The charcoal is packed into 7-cm-long by 6-mm o.d., 4-mm i.d. glass tubes and sealed at both ends. The packing consists of two sections, 100 mg in the absorbing section and 50 mg in the backup section, separated by 2 mm of urethane foam. The absorbing section is preceded by a plug of silylated glass wool, and the backup section is followed by 3 mm of urethane foam. The pressure drop across such a tube must be less than 1 inch of mercury at a flow rate of 1 L/min.*

Break the ends of the sampling tube to provide an opening of at least 2 mm and secure the tube in the sampling line. If tellurium containing particulates are suspected to be present, protect the charcoal tube with a membrane filter in a suitable cassette filter holder. To prevent channeling of the charcoal, keep the charcoal tube in a vertical position during sampling. A 400-L sample collected at a flow rate of 1 L/min is recommended. Record the ambient temperature and pressure of the atmosphere being sampled at 0, 3.5, and 7 h, and check the flow rate frequently during the sampling period. At the end of the sampling period, seal the tube with the plastic caps provided.

Transfer the charcoal from the absorbing and backup sections of the tubes to separate containers prefilled with 10.0 mL of 0.01 N sodium hydroxide solution. (The absorbing and backup sections are analyzed separately.) Stopper the containers and agitate gently for 60 min to desorb the tellurium.

Prepare tellurium standards in 0.01 N sodium hydroxide solution. Aspirate the samples and standards into the oxidizing air-acetylene flame and measure the absorbances at

---

*Two-stage coconut charcoal absorption tubes are available from SKC, Inc., Eighty Four, PA.

214.3 nm. Prepare a calibration curve from the absorbances of the standards and determine the tellurium content of the samples by direct comparison.

It is necessary to determine the desorption efficiency for the tellurium hexafluoride from the charcoal. This can be accomplished by injecting a known amount of the compound (using a gas-tight syringe) into a charcoal tube and establishing a recovery factor by determination of tellurium in the tube.

The absorption efficiency of the charcoal tubes is monitored by measuring the tellurium contents of the backup sections of the tubes. The absorption efficiency is considered to be 100% when no tellurium is found in the backup sections.

4.27. Procedure for Thallium:
NIOSH Method S306

Procedure S306 describes the collection, dissolution, and determination of thallium in the airborne particulate material of the occupational environment. The samples are collected on membrane filters, the matrices are destroyed and the thallium solubilized by nitric acid digestion, and the thallium contents of the samples are determined by conventional flame atomic absorption spectrometry. This method was validated over the range of 0.034 mg/m^3 to 0.15 mg/m^3 using 540-L samples. On this basis the working range of the method is estimated to be from 0.02 mg/m^3 to 0.25 mg/m^3. The coefficient of variation for the total analytical and sampling method was 6%. The collection efficiency and recovery were 99.1% and 99.9%, respectively.

To collect particulate thallium, use a personal sampling pump to pull air through a 37-mm diameter, 0.8-μm pore size mixed cellulose ester membrane filter contained in a three-piece cassette filter holder. Collect a sample equivalent to 540-L at a flow rate of 1.5 L/min. After sampling, replace the plugs in the cassette.

Transfer the filters containing the samples as well as filters spiked at 0.5, 1, and 2 times the OSHA standard to separate 125-mL beakers. Add to each beaker 2 mL of concentrated nitric acid, cover with watch glasses, and heat gently until the filters are dissolved and the volumes are reduced to

0.5 mL. Repeat the addition of acid and heating twice. Cool the solutions, add 10 mL of distilled water, and quantitatively transfer to 25-mL volumetric flasks. Bring the contents of the flasks to volume with distilled water.

Prepare the atomic absorption spectrometer with a thallium electrodeless discharge lamp. Aspirate the standards, the solutions from the sample filters, and those from the spiked filters into the oxidizing air-acetylene flame and measure the corresponding absorbances at 276.8 nm. Prepare a calibration curve from the absorbances of the standards and determine the thallium recovery factor and the thallium contents of the samples by direct comparison.

4.28. Procedure for Tin (Inorganic Compounds): NIOSH Method S183

Procedure S183 describes the collection, dissolution, and determination of tin in airborne particulates from the work place. The sample is collected on a membrane filter, the matrix is destroyed by nitric-sulfuric acid digestion, the tin is solubilized in hydrochloric acid, and the tin is determined by conventional flame atomic absorption spectrometry. This method was validated over the range of 0.939 mg/m^3 to 4.32 mg/m^3 using a 250-L sample. The working range of the method is estimated as 0.5 mg/m^3 to 6 mg/m^3. The total coefficient of variation for both the analytical and the sampling aspects of the method was 5.5%, and the collection efficiency was found to be 100%.

Collect the sample on a 37-mm diameter, 0.8-μm pore size mixed cellulose ester membrane filter contained in a three-piece cassette filter holder with a cellulose backup pad. Remove the cassette plugs and connect the sampling pump tubing. Clip the cassette to the worker's lapel. A sample size of 250 L is recommended. Sample at a flow rate of 1.5 L/min. Check the flow rate frequently during the sampling period. Terminate sampling at the predetermined time and note the sample flow rate, collection time, and ambient temperature and pressure.

Transfer the filters containing the samples and filters spiked at 0.5, 1, and 2 times the OSHA standard to separate 125-mL beakers. Treat the contents of each beaker with 1 mL

of concentrated sulfuric acid and 5 mL of concentrated nitric
acid, cover with watch glasses, and heat on a low-temperature
hotplate until fumes of SO_3 appear. Cool, and wash down the
watch glasses and insides of the beakers with distilled water.
Again heat the beakers until dense white fumes of SO_3 appear.
Transfer the beakers to a 200°C muffle furnace for 1 h to evaporate to dryness. Cool the beakers and add exactly 5 mL of
10% (v/v) hydrochloric acid to each. Mix well to assure complete solubilization of the tin.

Aspirate the solutions from the sample filters, the spiked
filters, and standards prepared in 10% hydrochloric acid into
the reducing air-acetylene flame. Measure the absorbances at
224.6 nm. Prepare a calibration curve from the absorbances of
the standards and determine the tin contents of the samples
and the recovery factors by direct comparison.

4.29. Procedure for Titanium (Dioxide):
NIOSH Method S385

Procedure S385 describes the collection, dissolution, and
determination of titanium in airborne particulates from the
work place. The sample is collected on a membrane filter, the
matrix is destroyed by nitric acid digestion, the titanium is
dissolved in sulfuric acid, and the titanium is determined by
conventional atomic absorption spectrometry in the nitrous
oxide-acetylene flame. This method has been validated over
the range of 8.1 mg/m^3 to 29.5 mg/m^3 using a 100-L sample.
The working range of the method is estimated to be from
2 mg/m^3 to 30 mg/m^3. The total coefficient of variation for
sampling and analysis was 11.2%. The collection efficiency
and recovery were 100% and 97%, respectively.

Collect the sample on a 37-mm diameter, 0.8-μm pore size
mixed cellulose ester membrane filter secured in a three-piece
cassette filter holder. Sample at a rate of 1.5 L/min and
collect a 100-L sample. Record ambient conditions of temperature and pressure during the sampling period.

Carefully transfer the filters containing the samples to
125-mL beakers. Similarly, transfer the blank filter and
filters spiked at 0.5, 1, and 2 times the OSHA standard to
125-mL beakers. Treat the contents of each beaker with 3 mL
of concentrated nitric acid, cover with watch glasses, and

heat on a low-temperature hotplate until the filters are dissolved and the volumes are reduced to 0.5 mL. Repeat this process once using 3 mL more of nitric acid. Cool and add to the contents of each beaker 8 mL of sulfuric acid-ammonium sulfate mixture (dissolve 40 g of ammonium sulfate in 100 mL of sulfuric acid). Add a few glass beads to each beaker and heat on a high-temperature hotplate for 60 min to assure complete solution. Do not allow the solutions to go to dryness. Cool and dilute with 10 mL of distilled water. Quantitatively transfer the contents of the beakers to 50-mL volumetric flasks. If fluorides are known to be present, add 5 mL of 1 N ammonium fluoride to each flask. Bring to volume with distilled water.

Prepare titanium standards in the sulfuric acid-ammonium sulfate matrix. If the ammonium fluoride was added to the sample solutions, add it also to the standards. Aspirate sample solutions and standards under identical conditions into the reducing nitrous oxide-acetylene flame. Measure the absorbances at 364.3 nm. Construct a calibration curve from the absorbances of the standards and determine the recovery factors and the titanium content of the samples by direct comparison.

4.30. Procedure for Tungsten (Particulate and Compounds): NIOSH Method P&CAM 271

Procedure P&CAM 271 describes the collection, selective dissolutions, and separate determinations of soluble and insoluble tungsten in airborne particulate matter collected from the occupational environment. The sample is collected on a membrane filter. Soluble tungsten is extracted from the filters with water. Interfering materials are removed by extracting the filter with dilute hydrochloric acid. Insoluble tungsten is dissolved by digesting the filter in nitric-hydrofluoric acid mixture and solubilizing the residue in sodium hydroxide solution. Soluble and insoluble tungsten are determined in their respective fractions by atomic absorption spectrometry using the nitrous oxide-acetylene flame. For soluble tungsten, the working range is from 0.14 mg/m^3 to 6.9 mg/m^3. The working range for insoluble tungsten is from 0.35 mg/m^3 to 17.4 mg/m^3 using a 720-Liter sample. The precision of the analysis is reflected in a coefficient of variation of 2.9% obtained from the analysis of eight calibration standards (each measured six times) covering the concentration range of 10.2 µg/mL to 523 µg/mL. The recoveries of soluble tungsten from filters spiked with 360, 725, and 1451 µg were 99.6% ± 4.5%, 102% ± 2.0%, and 100% ± 1.2%, respectively. Recoveries of insoluble tungsten spiked with approximately 2400, 3600, and 5800 µg were 97.5% ± 2.7%, 100% ± 4.3%, and 103% ± 2.8%, respectively.

Collect the sample on a cellulose ester membrane filter, 0.8-µm pore size, 37-mm diameter Millipore type AA or equivalent, secured in a Millipore MAWP 037 AO filter holder, or equivalent. Sample at a flow rate of 1.5 L/min for 8 h to collect a 720-L sample. Check the flow rate frequently during the sampling period. Measure and record the temperature and pressure of the atmosphere being sampled.

Remove the filters from the cassettes and carefully place them, sample side up, in Büchner funnels on side arm flasks or test tubes. Apply a few drops of water to the funnel and apply the vacuum to firmly seal the filter. Release the vacuum, add approximately 3 mL of water to the funnel, and allow to stand for 2 min. Apply the vacuum. Release the vacuum and add 3 mL of water to the funnel. Allow to stand for 2 min and reapply the vacuum. Transfer the filtrates to 10 mL volumetric flasks, rinse the side arm flasks or test tubes with 1 mL of 20% (w/v) sodium sulfate solution followed by 1 mL of water, and add the rinses to the appropriate volumetric flasks. Bring the contents of the volumetric flasks to volume with distilled water and identify as soluble tungsten.

Transfer the filters from the Büchner funnels to 100 mL Teflon beakers. Add 4 ml of 1:1 hydrochloric acid to each beaker and heat for 30 min at 110°C. Cool and filter through unused type AA Millipore filters firmly seated in Büchner funnels on side arm flasks or test tubes. Rinse the original filters three times with 2 mL of 1:10 hydrochloric acid and twice with 2 mL of deionized water. The rinsings are poured through the corresponding filters on the Büchner funnels. The filtrates are discarded and the filters from the Büchner funnels are paired with the corresponding original filters in the Teflon® beakers.

To the contents of the Teflon® beakers, add 12 mL of concentrated nitric acid, cover the beakers with Teflon® covers, and

heat for 12 h at 200°C. Remove the covers and allow the contents of the beakers to go to dryness at 110°C. If charring
occurs, add 3 mL of nitric acid and again take to dryness.
Add 2 mL of nitric acid and 2 mL of hydrofluoric acid, swirl
to mix, and heat at 110°C for $2\frac{1}{2}$ h to evaporate the acids.
Add 2.5 mL of 0.5 M sodium hydroxide solution to the residues
in each beaker, and heat for 15 min at 110°C. Wash down the
beaker walls with 5 mL of deionized water and add 2.5 mL of
sodium sulfate solution (20%, w/v) to the contents of each
beaker. Quantitatively transfer the contents of each beaker
to 25-mL volumetric flasks and bring to volume with deionized
water. Identify these flasks as insoluble tungsten.

Prepare tungsten standards in the sodium hydroxide-sodium
sulfate matrix of the samples. Aspirate the samples and
standards into the nitrous oxide-acetylene flame and measure
the absorbances at 255.1 nm using a slit width of 0.2 nm.
Make absorbance measurements in triplicate and confirm the
calibration by measuring the absorbance of a midrange standard following every fifth sample. Prepare the calibration
curve from the absorbances of the standards and determine the
soluble and insoluble tungsten contents of the samples by
direct comparison.

4.31. Procedure for Vanadium (Oxide):
NIOSH Method P&CAM 290

Procedure P&CAM 290 describes the collection, dissolution,
and determination of vanadium pentoxide in the particulate
matter from the work place atmosphere. The sample is collected
on a membrane filter, the vanadium pentoxide is dissolved in
sodium hydroxide solution, and the vandium pentoxide content of
the sample is determined by furnace atomic absorption spectrometry. Based on a 25-L sample, the working range of this method is
from 0.2 mg/m^3 to 0.5 mg/m^3. The analysis of six replicates
of three filters spiked with 0.625, 1.25, or 2.5 μg of vanadium showed precision as a coefficient of variation of 1.9%.
The recovery of vanadium from these 18 filters was 101%.

Collect the sample on a 37-mm diameter, 0.8-μm pore size
mixed cellulose ester membrane filter contained, with a cellulose backup pad, in a two-piece cassette filter holder. Sample
for 15 min at a flow rate of 1.7 L/min to collect a 25-L

sample. Record the temperature and pressure of the atmosphere sampled.

Carefully transfer the filters containing the samples to 50-mL beakers, treat with 3 mL of 0.01 N sodium hydroxide solution, and heat in a 50°C water bath for 30 min. Cool the solutions and quantitatively transfer the clear liquid to 10 mL volumetric flasks containing 250 μL of concentrated nitric acid. Bring to volume with 0.01 N sodium hydroxide solution.

Prepare vanadium standards in 2% (v/v) nitric acid. Make triplicate 20-μL injections of samples and standards and measure the absorbance, corrected for background, at 318.4 nm using the following time-temperature program: dry, 20 s at 100°C; ash, 10 s at 1300°C, atomize, 8 s at 2800°C. Background correction is made with the deuterium arc lamp. Prepare a calibration curve from the absorbances of the vanadium standards and determine the vanadium content of the samples by direct comparison.

4.32. Procedure for Yttrium: NIOSH Method S200

Procedure S200 describes the collection, dissolution, and determination of yttrium in airborne particles from the work place. The sample is collected on membrane filter, the matrix is destroyed by nitric-perchloric acid digestion which also solubilizes the yttrium, and the yttrium is determined by atomic absorption spectrometry in the nitrous oxide-acetylene flame. The method was validated over the range of 0.529 mg/m^3 to 2.211 mg/m^3 using a 500-L sample. The working range of the method is from 0.15 mg/m^3 to 5 mg/m^3. The total coefficient of variation for sampling and analysis was 5.4%, and the collection efficiency was 100%.

The sample is collected on a 37-mm diameter, 0.8-μm pore size mixed cellulose ester membrane filter secured in a three-piece cassette filter holder with a cellulose backup pad. The filter cassette is connected to the personal sampling pump and clipped to the worker's lapel. A minimum sample of 500-L is recommended.

Sample at a flow rate of 1.5 L/min. Check the flow rate frequently during the sampling period and record the temperature and pressure of the atmosphere being sampled several times during the sampling period.

Transfer the filters containing the samples and filters spiked with yttrium at 0.5, 1, and 2 times the OSHA standard to 125-mL beakers. Add 5 mL of concentrated nitric acid to each beaker, cover with watch glasses, and heat on a low-temperature hotplate until most of the acid has evaporated. Add 2 mL more of nitric acid and 1 mL of perchloric acid, cover, and heat on a high-temperature hotplate until dense white fumes of perchloric acid are evolved. Cool and wash down the watch glasses and insides of the beakers with distilled water. Evaporate the solutions to dryness. Cool, and add exactly 5 mL of 0.6 M hydrochloric acid containing 1000 ppm sodium to dissolve the yttrium.

Aspirate the sample solutions, the solutions from the spiked filters, and the standards (made up in 0.6 M hydrochloric acid containing 1000 ppm sodium) into a reducing nitrous oxide-acetylene flame and measure the absorbances at 410.2 nm. Prepare the calibration curve from the absorbances of the standards and determine the recoveries and the yttrium content of the samples by direct comparison.

4.33. Procedure for Zirconium: NIOSH Method S185

Procedure S185 describes the collection, dissolution, and determination of zirconium in particulate material from the atmosphere of the occupational environment. The sample is collected on a membrane filter, the matrix is destroyed and the zirconium solubilized by digestion in nitric-perchloric acid mixture, and the zirconium is determined by atomic absorption spectrometry in the nitrous oxide-acetylene flame. This method was validated over the range of 2.14 mg/m^3 to 9.02 mg/m^3 using a 720-L sample. The working range of the method was estimated to be from 1 mg/m^3 to 20 mg/m^3. The total coefficient of variation for both analysis and sampling was 5.3%. The collection efficiency was 100% and the recovery was 102%.

Collect a 720-liter sample using a 37-mm diameter, 0.8-µm pore size mixed cellulose ester membrane filter in a three-piece cassette filter holder. Sample at a flow rate of 1.5 L/min. Check the flow rate frequently during the sampling period. Record the temperature and pressure of the atmosphere being sampled at 0, 4, and 8 hs.

Carefully transfer the filters containing the samples and

filters spiked at 0.5, 1, and 2 times the OSHA standard to
125-mL beakers. Add 2 mL of concentrated nitric acid to the
contents of each beaker, cover with watch glasses, and heat on
a low-temperature hotplate until the filters are dissolved and
the volumes are reduced to 0.5 mL. Repeat the addition of
nitric acid and again heat until the volume is reduced to 0.5
mL. Cool the content of the beakers and add 2 mL of nitric
acid and 1 mL of perchloric acid. Continue heating until
dense white fumes of perchloric acid are evolved. Do not allow
the evaporation to go to dryness. Cool the solutions and add
10 mL of high-purity water to each. Quantitatively transfer
the solutions to 25-mL volumetric flasks. Add 2 mL of 1.25 M
ammonium fluoride to each flask and bring to volume with distilled water.

Prepare the zirconium standards in 0.1 M nitric acid- 0.1 M
ammonium fluoride. Aspirate the solutions from the filters containing the samples, the solutions from the spiked filters,
and the standards into the nitrous oxide-acetylene flame and
measure the absorbances at 360.1 nm. Prepare a calibration
curve from the absorbances of the standards and determine the
zirconium contents of the samples and the recovery factors by
direct comparison.

In addition to the procedure cited above, the U.S. Environmental Protection Agency[11] has, under the authority of the
Clean Air Act, specified a reference method for the determination of lead in suspended particulate matter collected from
the ambient air. Although not of regulatory status, the American Public Health Association (APHA)[26] has published a book
of standard methods for sampling and analyzing ambient and
workplace air. Similarly, the American Society for Testing
and Materials (ASTM)[26] has published nonregulatory standards
which include methods for determining metals in the workplace
atmosphere by atomic absorption spectrometry. With few exceptions, these methods call for the collection of the sample on
an appropriate filter, dissolution or extraction of the analyte,
and determination by either flame or furnace atomic absorption
spectrometry. Varian[161] recently has made available a review
that describes the collection of samples on a porous graphite
tube which then becomes a part of the carbon rod analyzer.

5. METHODS FOR COMPLIANCE WATER
QUALITY MONITORING

The methods contained in this chapter are recommended or mandated for monitoring potable water supplies, surface and ground waters, and waste waters. The United States Environmental Protection Agency (EPA) has carefully selected methods to provide guidance for laboratories engaged in water quality monitoring. The Canadian Department of the Environment (DOE) has provided similar guidance. In both cases, the methods have undergone interlaboratory evaluations, and they have been found acceptable. There are instances, however, where a particular sample cannot be successfully evaluated by these procedures. The U.S. EPA's Methods for Chemical Analysis of Water and Wastes and the Canadian DOE's Analytical Methods Manual should be consulted if more details on these procedures are required. The procedures contained in the former are approved for monitoring ambient waters and waste waters under PL 92-500 and for potable water monitoring under PL 93-523.

5.1. Sample Preservation and Storage

With the exception of those in which mercury is to be determined, the U.S. EPA[9] allows samples for the determination of total metal to be kept in glass, polyethylene, or polypropylene containers for up to 6 months providing the pH of the sample has been adjusted to 2 or below with nitric acid. Samples for the determination of total mercury preserved by the addition of nitric acid to pH 2 or below may be kept in glass containers for 38 days or in polyethylene or polyproplyene containers for only 13 days prior to analysis.

The Canadian DOE[13] specifies a maximum storage time of 6 months for most metals in polyethylene containers preserved with the equivalent of 2 mL concentrated nitric acid per liter of sample. Exceptions are as follows:

Metal	Preservative	Storage time
Arsenic	Cool to 4°C	6 months
Boron	Cool to 4°C	6 months
Calcium	Cool to 4°C	7 days
Magnesium	Cool to 4°C	7 days
Mercury	1 mL concentrated sulfuric acid and 1 mL 5% potassium dichromate per 100-mL sample	1 month in glass or Teflon® containers
Potassium	Cool to 4°C	7 days
Selenium	Cool to 4°C	6 months
Silver	0.4 g disodium EDTA per 100-mL sample	10 days
Sodium	Cool to 4°C	7 days

5.2. General Procedures for the Determination of Metals

Both the U.S. EPA and the Canadian DOE have developed procedures for the determination of dissolved, suspended, extractable, and total metal levels in water samples, and they have similarly developed general and specific procedures for the treatment of samples prior to the determination of trace metals in samples of industrial waste.

5.2.1. U.S. EPA Procedures

Preliminary treatment of waste water and/or industrial effluents is usually necessary because of the complexity and variability of the sample matrix. Suspended material must be subjected to a solubilization process before analysis. This process may vary because of the metals to be determined and the nature of the sample being analyzed. When the decomposition of organic material is necessary, the process should include a wet digestion with nitric acid.

In those instances where complete characterization of a sample is desired, the suspended material must be analyzed separately. This may be accomplished by filtration and acid digestion of the suspended material. Metallic constituents in this acid digest are subsequently determined and the sum of the dissolved plus the suspended concentrations will then provide the total concentration present. The sample should be filtered as soon as possible after collection and the filtrate acidified immediately.

The total sample may also be treated with acid without prior filtration to measure what may be termed total recoverable concentrations.

For the determination of dissolved constituents, the sample must be filtered through a 0.45 μm membrane filter as soon as possible after collection. The filtrate should be acidified with 1:1 redistilled nitric acid to a pH of 2 or less. If hexavalent chromium is to be determined, a portion of the filtrate should be transferred to a separate container prior to acidification and analyzed as soon as possible. The results of metal determinations in the filtrate are identified as dissolved concentrations.

For the determination of suspended metals, a measured volume of the unpreserved sample is filtered through a 0.45-μm membrane filter. The filter containing the suspended material is transferred to a 250-mL beaker and treated with 3 mL of concentrated, redistilled nitric acid. The beaker is covered with a watch glass and heated gently. When the membrane filter has dissolved, increase the heat and continue heating until the volume of acid is greatly reduced. Add an additional 3 mL of concentrated, redistilled nitric acid and continue heating to complete the digestion as indicated by a light color of the solution. Evaporate to near dryness, add 5 mL of 1:1 redistilled hydrochloric acid, and heat to dissolve any soluble material. If furnace techniques are to be employed, substitute 1 mL of 1:1 redistilled nitric acid for the hydrochloric acid. Wash down the watch glass and walls of the beaker with deionized water and filter the solution to remove insoluble material. Transfer the filtrate to a volumetric flask and dilute to the mark. The results of metal determinations in the digested sample filtrate are identified as suspended concentrations.

For the determination of total metals, a measured volume of the well-mixed, acidified, but unfiltered sample is transferred to a beaker, treated with 3 mL of concentrated, redistilled nitric acid, and heated to near dryness. An additional

3 mL of redistilled nitric acid are added, the beaker is covered with a watch glass, and the contents are heated gently
until digestion is complete. The solution is again evaporated
to near dryness and treated with 5 mL of 1:1 hydrochloric acid.
The solution is heated to dissolve the soluble material.

Use 1 mL of 1:1 redistilled nitric acid instead of hydrochloric acid if furnace techniques are to be employed. Wash
down the beaker walls and watch glass with deionized water,
filter to remove insoluble material, transfer to a volumetric
flask, and dilute with deionized water. The results obtained
from samples prepared in this manner are identified as total
concentrations.

To determine total recoverable metals, acidify the entire
sample at the time of collection with an amount of concentrated redistilled nitric acid equivalent to 5 mL/L of sample.
Transfer a 100-mL aliquot to a beaker or flask, add 5 mL of
1:1 redistilled hydrochloric acid, and heat gently until the
volume is reduced to 15 or 20 mL. Omit the hydrochloric acid
if the sample is to be analyzed by furnace techniques. Filter
the sample, transfer to a 100-mL volumetric flask, and dilute
with deionized water. If the total dissolved solids of the
original sample do not exceed 500 mg/L, if background corrections are applied, and if there are no losses by precipitation,
the dilution may be made in a volumetric flask of lesser capacity, thereby effecting a corresponding increase in the
concentration of analyte in the prepared sample. Results
obtained from samples prepared in this manner are identified
as total recoverable concentrations.

When the concentration of metal is not sufficiently high
to determine directly, or when considerable dissolved solids
are present in the sample, some of the metals may be chelated
and extracted with organic solvents. This approach serves to
increase the concentration of analyte in the prepared sample
and, simultaneously, remove matrix interferences. Ammonium
pyrrolidine dithiocarbamate (APDC) in methyl isobutyl ketone
(MIBK) is widely used for this purpose. It is particularly
useful for zinc, cadmium, iron, manganese, copper, silver,
lead, and hexavalent chromium. The prescribed procedure is
described in section 3.2.1.

5.2.2. Canadian DOE Procedures

To determine the dissolved metal concentration of the

sample, the sample taken for analysis should be free from tur-
bidity, or it should be filtered through a 0.45-μm membrane
filter. The sample should be filtered at the time of collection, and it should be treated with the equivalent of 2 mL
concentrated nitric acid per liter of sample after filtration.

For the determination of suspended metal concentrations,
filter a measured volume of sample through a 0.45-μm membrane
filter. Transfer the filter and its contents to a 250-mL
beaker, treated with 3 mL of concentrated nitric acid, and
heat gently. When the membrane has dissolved, increase the
heat and evaporate the solution to dryness. Cool and add
another 3 mL of concentrated nitric acid. Continue heating
until digestion is complete. Evaporate to dryness. Add 2 mL
of 1:1 hydrochloric acid to the dry residue and heat to dissolve the residue. Wash down the sides of the beaker with
deionized water and filter, if necessary, to remove insoluble
material. Dilute this solution with deionized water to the
original volume of sample measured out.

The total metal concentration of the sample is the sum of
the dissolved and suspended metal concentrations.

The extractable metal content can be determined by analysis
after mineral acid has been added to the sample bottle and the
sample has been shaken and allowed to stand overnight. In
many cases, this represents the total metal content if all
sediment has dissolved.

If the concentration of metal is not sufficient to be determined by direct aspiration, or when the sample contains
large amounts of dissolved solids, some metals may be chelated
and extracted with organic solvents. The procedure for the
determination of antimony, cadmium, cobalt, copper, iron, lead,
nickel, and zinc is outlined below.

1. Pipette 100 mL of sample into a 250-mL volumetric flask.

2. Add 5.0 mL of buffer made by dissolving 272 g of sodium
   acetate in a liter of distilled water, adding acetic
   acid until a pH of 4.75 is obtained, and diluting to a
   final volume of 2 L. The buffer should be extracted
   with 0.01% dithizone in carbon tetrachloride to remove
   impurities prior to use.

3. Add 5.0 mL of freshly prepared 1% APDC. Mix by swirling.

4. Add 10.0 mL of MIBK and shake the flasks for 5 min.

5. Permit the layers to separate and slowly add deionized water down the side of the flask to float the organic layer up to the top of the flask.

6. Aspirate the ketone layer and record the absorbance for each sample and standard.

5.3. Procedures for Aluminum

Aluminum is conveniently determined in a rich nitrous oxide-acetylene flame at a wavelength of 309.3 nm. Alternate wavelengths are 308.2 nm, 396.2 nm, and 394.4 nm. Ionization interferences are corrected for by the addition of potassium chloride. The thermal stability of most aluminum compounds allows ashing temperatures of up to 1400°C for the furnace technique. Few interferences are encountered, but the presence of chloride and the use of nitrogen purge gas should be avoided.

5.3.1. U.S. EPA Methods 202.1 and 202.2 for Aluminum

Method 202.1 is applicable to the determination of aluminum in the concentration range of from 5 mg/L to 50 mg/L by direct aspiration of prepared wastewater samples. The samples are prepared by the procedures described in Section 5.2.1, and the prepared samples and standards are treated with potassium chloride solution (95 g/L) at a rate of 2 mL per 100 ml of sample or standard. The treated samples and standards are aspirated into the nitrous oxide-acetylene flame, and the absorbances at 309.3 nm are measured. The absorbances of the standards are used to establish the calibration curve, and the aluminum content of the samples is determined by direct comparison.

Natural water samples spiked with six concentrates containing varying amounts of aluminum, cadmium, chromium, iron, manganese, lead, and zinc were analyzed by some three dozen laboratories. Mean aluminum levels were within 10% of theoretical in the concentration range of 600 ppb to 1200 ppb. At lower aluminum levels, 15 ppb to 35 ppb, pronounced positive biases were observed.

Method 202.2 describes the furnace technique for the determination of aluminum in the concentration range of 20 µg/L

to 200 µg/L. The samples are prepared by the procedures in
Section 5.2.1. For every sample matrix analyzed, it is necessary to show that the method of standard additions is not
needed if the aluminum contents of the sample is to be determined by direct comparison. The prepared samples and samples
spiked with known amounts of aluminum are injected into the
graphite furnace, and replicate measurements of the absorbance
are made at 309.3 nm using the following time-temperature
program: dry, 30 s at 125°C; ash, 30 s at 1300°C; atomize,
10 s at 2700°C. Argon should be used as the purge gas.
Aluminum forms a stable nitride.

5.3.2. Canadian DOE Methods 13002 and 13003 for Aluminum

Method 13001 is applicable to the determination of aluminum in a wide variety of waters including surface waters,
domestic water, industrial waste waters and sea water. The
samples are prepared by the procedures described in Section
5.2.2, and the prepared samples and standards in the same
acid matrix are aspirated into the nitrous oxide-acetylene
flame. The absorbances are measured at 309.3 nm, and the
aluminum content of the samples is determined by direct comparison using the absorbances of the standards for the
calibration curve.

Method 13002 is a chelation-extraction procedure for the
determination of low levels of aluminum. The samples are
prepared by the procedures in Section 5.2.2. Both the samples and standards are extracted as follows:

Pipette 100-mL aliquots of the prepared samples and standards into 250-mL separatory funnels, add 2 mL of 2% 8-hydroxy-
quinoline in 1 N acetic acid, and add 3 mL of buffer prepared
by dissolving 200 g of ammonium acetate in water containing
70 mL of concentrated ammonia and diluting to 1 L with deionized water. Check to determine whether a pH of 8 has been
established. If not, adjust to pH 8. Add 5 mL of chloroform
and shake the funnel vigorously. Allow the phases to separate
and drain the lower chloroform layer into a test tube. Repeat
the extraction with a second 5-mL portion of chloroform and
add the lower layer to the contents of the test tube. The
chloroform extracts are aspirated into the nitrous oxide-
acetylene flame, and the absorbances at 309.3 nm are recorded.
The calibration curve is prepared from the absorbances of the

standards, and the aluminum content of the samples is determined by direct comparison.

5.4. Procedures for Antimony

The determination of antimony is made at a wavelength of
217.6 nm in the air-acetylene flame. Because of its greater
light output and longer life, the electrodeless discharge
lamp (EDL) is preferred to the hollow cathode lamp (HCL).
Possible spectral interference from lead can be avoided by
measuring the antimony absorbance at an alternate wavelength
of 231.1 nm. The regulatory agencies have not recommended the
argon-hydrogen flame or the hydride generation technique for
the determination of antimony.

5.4.1. U.S. EPA Methods 204.1 and 204.2 for Antimony

Method 204.1 makes use of the air-acetylene flame for the
determination of antimony at levels of from 1 to 40 mg/L in
wastewater samples prepared by the nitric acid digestion described in Section 5.2.1. Samples and standards prepared in
nitric acid are aspirated into a lean air-acetylene flame,
and the absorbances are measured at 217.6 nm. The antimony
levels of the samples are determined by direct comparison.
The precision of the method was evaluated using a mixed domestic-industrial effluent spiked at 5 and 15 mg/L. The coefficients of variations were 8% and 1%, respectively, and the
recoveries were 96% and 97%, respectively.

Method 204.2 is directed to the determination of antimony
by the furnace technique. The optimum concentration range
for this method is from 20 to 300 µg/L. The wastewater samples are prepared by the nitric digestion and dissolved in
5 mL of 1:1 hydrochloric acid. The final solutions as well
as the standards should be made up in 2% (v/v) nitric acid.
The absorbances of replicate 20 µL injections are measured
at 217.6 nm. The following time-temperature program is recommended: dry, 30 s at 125°C; ash, 30 s at 800°C; atomize,
10 s at 2700°C. Background correction is recommended. The
purge gas is nitrogen on continuous flow. The antimony content of the samples is determined by the method of standard
additions. Direct comparison may be used if the absence of

matrix effects can be demonstrated.

5.4.2. Canadian DOE Methods 51001 and 51002 for Antimony

Method 51001 parallels the U.S. EPA Method 204.1.

Method 51002 makes use of the solvent extraction procedure described in Section 5.2.2 to concentrate the antimony prior to aspiration of the organic phase into the flame and measurement of absorbance at 217.6 nm. The determination of antimony is by direct comparison using standards extracted and measured under identical conditions.

5.5. Procedures for Arsenic

Samples for the determination of arsenic are usually prepared by acid digestion in the presence of strong oxidizing agents. Absorbance measurements are made at a wavelength of 193.7 nm. The electrodeless discharge lamp is preferred over the hollow cathode lamp. The arsenic is converted to the hydride prior to atomization, which is achieved by thermal decomposition of the hydride in either the argon-hydrogen flame or a heated absorption cell. Direct aspiration is not recommended by the regulatory agencies. The U.S. EPA does, however, recommend the furnace technique for the determination of arsenic.

5.5.1. U.S. EPA Methods 206.2, 206.3 and 206.5 for Arsenic

Method 206.5 describes the preparation of water and waste-water samples for the determination of arsenic by the hydride generation technique. To a well-mixed 50-mL sample, add 7 mL of 1:1 sulfuric acid and 5 mL of concentrated nitric acid. Cautiously heat the acidified sample until dense white fumes of SO_3 are evolved from a colorless or pale straw solution. If charring occurs, stop the digestion immediately, cool, and add an additional 5 mL of nitric acid before continuing. The colorless or pale straw solution is cooled, treated with 25 mL of distilled water, and again evaporated to dense white fumes of SO_3. Cool, quantitatively transfer to a 50-mL volumetric flask, add 20 mL of concentrated hydrochloric acid, and bring to volume with distilled water.

Method 206.3 describes the determination of arsenic in prepared samples by the hydride generation technique. A 25-mL aliquot of the prepared sample solution or standard is transferred to the reaction vessel of the hydride generator. Add 1 mL of 20% (w/v) potassium iodide solution and 0.5 mL of stannous chloride solution (prepared by dissolving 100 g of stannous chloride in 100 mL of concentrated hydrochloric acid) to reduce the arsenic to the trivalent state. Attach the reaction vessel to the argon gas flushing system and inject 1.5 mL of zinc slurry into the air-tight reaction vessel. The trivalent arsenic is reduced to arsine and flushed into the argon-hydrogen flame of the atomic absorption spectrometer. Record the absorbance at 193.7 on a strip chart. The arsenic content of the sample is determined by direct comparison. The sample should be split and spiked to establish freedom from interferences and quantitative recoveries. Ten replicate solutions of o-arsenilic acid at concentrations of 5, 10, and 20 µg/L were analyzed by this method. The coefficients of variation were 6%, 9%, and 5%, respectively, and the recoveries of arsenic were 94%, 93%, and 85%, respectively.

Method 206.2 describes the sample preparation and furnace atomic absorption technique for the determination of arsenic in water and wastewater samples.

Transfer 100 mL of well-mixed sample to a 250-mL beaker and 2 mL of 30% hydrogen peroxide and 1 mL of concentrated nitric acid. Heat for 1 h or until the volume is reduced to slightly less than 50 mL. Cool and bring to a total volume of 50 mL. Pipette 5 mL of this solution into a 10 mL volumetric flask, add 1 mL of 1% w/v nickel nitrate solution, and bring to volume with high-purity water.

Replicate 20-µL aliquots of the sample and standards (prepared with appropriate amounts of hydrogen peroxide, nitric acid, and nickel nitrate) are injected into the furnace, and the absorbances are measured at 193.7 nm using a strip chart recorder. The following time-temperature program is recommended: dry, 30 s at 125°C; ash, 30 s at 1100°C; atomize, 10 s at 2700°C. Background correction is recommended. Unless it can be demonstrated that there are no matrix effects, the method of standard additions is to be used for the determination of arsenic. Direct comparison may be used when matrix interferences are shown to be absent.

Tap water samples spiked with 20, 50, and 100 µg/L of

arsenic were analyzed by this method. Replicate analysis
showed coefficients of variation of 3.5%, 2.2%, and 1.6%,
respectively. Recoveries of the spikes at these levels were
105%, 106%, and 101%, respectively. A mixed industrial-domestic effluent containing 15 µg/L of arsenic was spiked with 2,
10, and 25 µg/L additional arsenic. Recoveries were 85%, 90%,
and 88% at these levels. The coefficients of variation were
8.8%, 8.2%, 5.4%, and 8.7% for the samples containing 15, 17,
25, and 40 µg/L, respectively.

5.5.2. Canadian DOE Method 33304 for Arsenic

Method 33304 describes the sample preparation procedures
and the flameless atomic absorption techniques for the determination of arsenic in surface water, ground water, saline
water and industrial waste water. The sample is digested in
acid persulfate; the arsenic is liberated from the solution
as the gaseous hydride and passed to a heated, open-ended tube
furnace where the arsenic is atomized; and the absorbance is
measured at a wavelength of 193.7 nm. The working range of
the method is from 0.1 µg/L to 50 µg/L.

To 50-mL of sample add 5 mL of 2% w/v potassium persulfate
solution and 1 mL of concentrated hydrochloric acid; boil
vigorously for at least 15 min, cool, and dilute to 50 mL.
The prepared samples and standards are then treated, sequentially, in an automated system, with stannous chloride-
hydrochloric acid solution, potassium iodide solution, and
aluminum slurry for the generation of arsine which is flushed
into the open-ended tube furnace in the optical system of the
atomic absorption spectrometer. The manifold for the determination of arsenic is shown in Figure 5.1.

In a single laboratory, the coefficients of variation at
arsenic levels of 1.8 µg/L, 2.1 ug/L, and 4.4 µg/L were 4.8%,
4.2%, and 3.2%, respectively. The recoveries of 0.5 µg/L,
1.0 µg/L, and 3.0 µg/L arsenic spikes were 80%, 81%, and
103%, respectively.

5.6. Procedures for Barium

Barium is conveniently determined in a rich nitrous oxide-
acetylene flame at a wavelength of 533.6 nm. The determination

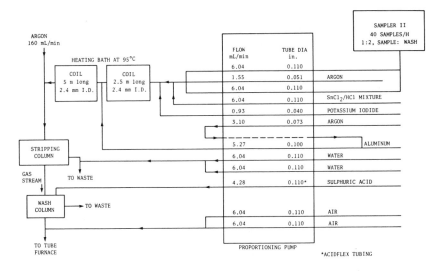

Figure 5.1. Arsenic manifold.[13]

is subject to ionization interference. Hence, sodium or potassium is added to both samples and standards.

5.6.1. U.S. EPA Methods 208.1 and 208.2 for Barium

Method 208.1 describes the preparation of samples and the
conventional flame atomic absorption techniques for the determination of barium in potable waters and waste waters. The
samples are prepared for analysis by the procedures described
in Section 5.2.1. To each 100 mL of sample and standard is
added 2.0 mL of potassium chloride solution (95 g/L). Samples
and standards are aspirated into the nitrous oxide-acetylene
flame, and the absorbances are measured at 553.6 nm. A calibration curve is prepared from the absorbances of the standards and the barium content of the samples is determined by
direct comparison.

Three samples containing barium at 500 µg/L, 1000 µg/L,
and 5000 µg/L were analyzed by 13 laboratories. The coefficients of variation were 10%, 8.9%, and 3.7%, respectively,
and the relative errors at these concentrations were 8.6%,
2.7%, and 1.4%, respectively. A single laboratory using a
mixed domestic-industrial effluent spiked at 0.40 mg/L and
2.0 mg/L barium showed precisions corresponding to

coefficients of variation of 11% and 6.5%, respectively.

Method 208.2 describes the determination of barium by furnace atomization. The samples are prepared for analysis in accord with the procedures described in Section 5.2.1. Replicate 20-µL injections of samples and standards in 0.5% (v/v) nitric acid are made, and the absorbances are measured at 553.6 nm. The following program of time and temperature is recommended: dry, 30 s at 125°C; ash, 30 s at 1200°C; atomize, 10 s at 2800°C. Continuous flow argon purge gas and pyrolytic graphite are recommended. The determination of barium may be made by direct comparison if the absence of matrix interferences can be demonstrated; otherwise, the method of standard additions is required.

Tap water samples spiked at 500 µg/L and 1000 µg/L barium were analyzed with precisions corresponding to coefficients of variation of 0.5% and 0.2%, respectively. Recoveries at these levels were 96% and 102%, respectively.

5.6.2. Canadian DOE Method 56001 for Barium

This method is applicable to the determination of barium in surface waters, domestic waters, industrial waste waters, and sea water. The samples are prepared for analysis by the procedures described in Section 5.2.2.

To 100 mL of the prepared sample or standard, add 2 mL of sodium chloride solution (250 g/L). Aspirate the samples and the standards into the nitrous oxide-acetylene flame and measure the absorbances at 553.6 nm. Prepare a calibration curve from the absorbances of the standards and determine the barium levels of the samples by direct comparison.

5.7. Procedures for Beryllium

The determination of beryllium is made at a wavelength of 234.9 nm. A rich nitrous oxide-acetylene flame is recommended.

5.7.1. U.S. EPA Methods 210.1 and 210.2 for Beryllium

Method 210.1 describes the sample preparation procedures and conventional flame atomic absorption techniques for the determination of beryllium in waste waters. The samples are prepared by the procedures described in Section 5.2.1. Samples and standards are aspirated in to the nitrous oxide-acetylene flame, and the absorbances are measured at 234.9 nm. The beryllium content of the samples is determined by direct comparison.

Using a mixed domestic-industrial effluent spiked at beryllium levels of 0.01, 0.05, and 0.25 mg/L, a single laboratory demonstrated precision corresponding to coefficients of variation equivalent to 10%, 2%, and 0.8%, respectively. Recoveries at these levels were: 100%, 98%, 97%, respectively.

Method 210.2 describes the determination of beryllium by the furnace technique after the samples have been prepared by the procedures of Section 5.2.1. The samples and standards should be made up in 0.5% v/v nitric acid. Replicate 20-µL aliquots of the standards and samples are injected into the furnace, and the absorbances are measured at 234.9 nm. The following time-temperature program is recommended: dry, 30 s at 125°C; ash, 30 s at 1000°C; atomize, 10 s at 2800°C. Continuous flow argon purge gas and nonpyrolytic graphite are recommended.

5.7.2. Canadian DOE Methods 04001 and 04002 for Beryllium

Method 04001 describes the determination of beryllium in samples of surface waters and industrial waste waters prepared for analysis by the procedures in Section 5.2.2. The determinations are made by conventional atomic absorption spectrometry in the nitrous oxide-acetylene flame at a wavelength of 234.8 nm. Direct comparison is employed.

Method 04002 is a chelation-extraction procedure for determining low levels of beryllium in samples prepared in accord with Section 5.2.2. The chelation-extraction is performed with 8-hydroxyquinoline and chloroform using the same procedures described for aluminum in Section 5.3.2. The absorbance measurements are made at 234.9 nm.

5.8. Procedures for Cadmium

Cadmium is conveniently determined in the air-acetylene flame at a wavelength of 228.8 nm. The furnace determination may suffer loss of cadmium by premature volatilization. The cadmium is rendered more refractory by the addition of phosphate to both samples and standards.

5.8.1. U.S. EPA Methods 213.1 and 213.2 for Cadmium

Method 213.1 describes the preparation procedures and atomic absorption techniques for the determination of cadmium in potable waters and waste waters. Samples are prepared by the procedures described in Section 5.2.1. For prepared samples containing less than 20 μg/L, or for samples with high dissolved solids content, the chelation-extraction procedure described in Section 3.2.1. can be applied. Samples and standards are aspirated into the air-acetylene flame under identical conditions, and their absorbances are measured at a wavelength of 228.8 nm. The calibration curve is prepared from the absorbances of the standards, and the cadmium content of the samples is determined by direct comparison.

Six synthetic concentrates containing varying amounts of aluminum, cadmium, chromium, copper, iron, manganese, lead, and zinc were used to spike natural water. Some 65 laboratories analyzed these spiked water samples. For those spiked in the range of 75 ppb cadmium, the mean value of the participating laboratories showed a 5% negative bias. The bias was approximately 20% positive for samples spiked at 15 ppb.

Method 213.2 describes the determination of cadmium by furnace atomic absorption spectrometry. To 100 mL of prepared sample or standard in 0.5% (v/v) nitric acid, add 2 mL of 40% (w/v) diammonium monohydrogen phosphate solution. Inject replicate 20-μg/L aliquots into the graphite furnace programmed for the following time-temperature cycle: dry, 30 s at 125°C; ash, 30 s at 500°C; atomize, 10 s at 1900°C. The use of continuous argon purge and nonpyrolytic graphite tubes is recommended. Absorbance measurements are made at 228.8 nm. For every sample matrix analyzed, it is necessary to verify that the method of standard additions is not necessary. If this is not done, the cadmium content of the samples must be determined by the method of standard additions.

A tap water sample spiked with cadmium at levels of 2.5 μg/L, 5 μg/L, and 10 μg/L showed coefficients of variation of 4%, 3%, and 3%, respectively, and the recoveries were 96%, 99%, and 98%, respectively.

5.8.2. Canadian DOE Methods 48001 and 48002 for Cadmium

Method 48001 describes the preparation procedures and the atomic absorption techniques for the determination of cadmium in a wide variety of waters, including those from the surface, the ground, and the sea. The samples are prepared as described in Section 5.2.2. The prepared samples and the standards are then aspirated into the air acetylene flame, and the absorbances are measured at 228.8 nm. The determination of cadmium is by direct comparison.

Method 48002 is the chelation-extraction procedure for the determination of cadmium at levels below those detectable by direct aspiration. Samples and standards are chelated and extracted in accord with the procedures in Section 5.2.2. The organic phases are aspirated into the air-acetylene flame, and the absorbances at 228.8 nm are recorded. The cadmium content of the samples is determined by direct comparison.

In a single laboratory, the coefficient of variation for the determination of cadmium at 10 µg/L by chelation-extraction was 3.5%.

5.9. Procedures for Calcium

Calcium absorbance is measured at a wavelength of 422.7 nm in either the air-acetylene flame or the nitrous oxide-acetylene flame. Chemical interferences in the former require the use of releasing agents, and ionization interferences in the later require the addition of deionizers.

5.9.1. U.S. EPA Method 215.2 for Calcium

This method describes the preparation procedures and the atomic absorption techniques for the determination of calcium in waste waters. The samples are prepared in accord with the procedures described in Section 5.2.1. To each prepared sample and standard in hydrochloric acid matrix is added lanthanum chloride solution (prepared by dissolving 29 g of lanthanum oxide in 250 mL of concentrated hydrochloric acid and diluting to 500 mL with high-purity water) in a ratio of 10 parts sample to 1 part lanthanum chloride solution. Samples and standards are aspirated into the air-acetylene flame, and

the absorbances are measured at 422.7 nm. The determination
of the calcium levels of the samples is made by direct
comparison.

The coefficients of variation for the determination of calcium in high-purity water at levels of 9 mg/L and 36 mg/L
were 3.3% and 1.5%, respectively. Recoveries at these levels
were 99% in both cases.

5.9.2. Canadian DOE Method 20103 for Calcium

Method 20103 describes the determination of calcium in a
wide variety of waters, including surface water, ground water,
and sea water. The samples are prepared by the procedures
described in Section 5.2.2. To 25 mL of prepared sample of
standard is added 0.5 mL of lanthanum chloride solution (prepared by dissolving 117.2 g of lanthanum chloride in a mixture of 600 mL of distilled water and 200 mL of concentrated
hydrochloric acid and diluting to 1 L). Samples and standards
are aspirated into the air-acetylene flame, and the absorbances are measured at 422.7 nm. The calcium levels of the
samples are determined by direct comparison.

In a single laboratory, the coefficient of variation at a
calcium level of 10 mg/L was 0.6%.

5.10. Procedures for Chromium

Chromium is most frequently determined at a wavelength of
357.9 nm. Alternate wavelengths are 359.4 nm, 360.5 nm,
425.4 nm, 427.5 nm, and 429.0 nm. Both the air-acetylene
flame and the nitrous oxide-acetylene flame are used. Sensitivity and selectivity depend upon the oxidant to fuel ratio,
and trivalent chromium is reported to show higher absorbances
than equal amounts of hexavalent chromium. Fewer interferences are experienced with electrothermal atomization. High
chloride levels, however, may lead to losses by premature
volatilization as chromyl chloride.

5.10.1. U.S. EPA Methods 218.1, 218.2, 218.3
and 218.4 for Chromium

Method 218.1 is applicable to the determination of chromium

in water and waste water. The working range of the method is
from 0.5 mg/L to 10 mg/L. Samples are prepared by the procedures described in Section 5.2.1. Samples and standards are
aspirated into a rich nitrous oxide-acetylene flame, and the
absorbances are measured at 357.9 nm. The chromium content
of the samples is determined by direct comparison.

Six concentrates containing varying amounts of aluminum,
cadmium, chromium, copper, iron, manganese, lead, and zinc
were used to spike natural water. The spiked samples were
analyzed for chromium by some six dozen laboratories. The
mean values for samples spiked at the 385-ppb level showed a
30% coefficient of variation and a 5% negative bias, and those
spiked at the 85-ppb level showed a 30% coefficient of variation and a 7% negative bias.

Method 218.2 describes the determination of water and waste
water chromium levels in the 5 µg/L to 100 µg/L range by furnace atomic absorption. Samples are prepared by the procedures described in Section 5.2.1. To each 100 mL of prepared
sample or standard in 0.5% nitric acid, add 1 mL of 30% hydrogen peroxide and 1 mL of calcium nitrate solution (prepared
by dissolving 11.8 g of calcium nitrate tetrahydrate in water
and diluting the resulting solution to 100 mL). The peroxide
is added to reduce the chromium to the trivalent state, and
the calcium is added to minimize variations due to its suppressive effects. Inject replicate 20-µL aliquots and measure the absorbances at 357.9 nm using the following time-
temperature program: dry, 30 s at 125°C; ash, 30 s at 1000°C;
atomize, 10 s at 2700°C. Background correction is necessary,
and the method of standard additions must be used unless it
has been demonstrated to be unnecessary.

A single laboratory using tap water spiked at chromium
levels of 19 µg/L, 48 µg/L, and 77 µg/L showed coefficients
of variation of 0.5%, 0.2%, and 1%, respectively. Recoveries
were 97%, 101%, and 102%, respectively.

Method 218.3 describes a chelation-extraction procedure for
the determination of chromium in water and waste water at
levels from 1 µg/L to 25 µg/L. To 100 mL of prepared sample
or trivalent chromium standard at pH 2 or below, contained in
a 200-mL volumetric flask, add 0.1 N potassium permanganate
dropwise until a faint pink color persists. Heat the samples
and standards on a steam bath for 20 min. If the pink color
disappears during this time, add additional potassium

permanganate to retain the color. After 20 min of heating,
while the flasks are still on the steam bath, add 0.1% (w/v)
sodium azide solution dropwise to destroy the faint pink color.
Continue heating for an additional 5 min. Cool the contents
of the flasks and filter through Whatman No. 40 paper or the
equivalent to remove the dark precipitate of oxides of manganese. Adjust the pH's of the filtrates to 2.4 by the addition
of 1 M sodium hydroxide solution and 0.12 M sulfuric acid.
Return the filtrates to the flasks. Add 5.0 mL of 1% w/v
ammonium pyrrolidine dithiocarbamate (APDC) solution and mix
well. Add 10.0 mL of methyl isobutyl ketone (MIBK) to the
flasks and shake them vigorously for 3 min. Allow the phases
to separate and add high-purity water down the sides of the
flasks to float the ketone layers up into the necks. Aspirate the ketone layers into a rich air-acetylene flame and
record the absorbances at 357.9 nm. This should be done in
replicate for each solution. Prepare a calibration curve
from the absorbances of the standards and determine the chromium content of the samples by direct comparison.

Method 218.4 is used for the selective determination of
hexavalent chromium in water samples. The working range of
this method is from 1 µg/L to 25 µg/L. The stability of
hexavalent chromium in water samples is unknown. Hence, un-
acidified samples or filtrates should undergo minimal storage
at 4°C prior to analysis.

Dilute blanks, aliquots of the samples, and standards,
contained in 200-mL volumetric flasks, to approximately 100
mL with high-purity water. Adjust the pHs of the contents of
the flasks to 2.4 with 1 M sodium hydroxide solution and 0.12
M sulfuric acid. Add 5.0 mL of 1% w/v ammonium pyrrolidine
dithiocarbamate solution and mix well. Add 10.0 mL of methyl
isobutyl ketone to the flasks and shake them for 3 min.
Allow the layers to separate and add high-purity water down
the sides of the flasks until the ketone layers are floated
up into the necks of the flasks. Aspirate the ketone layers
into a rich air-acetylene flame and make replicate measurements
of the absorbances at 357.9 nm. Prepare a calibration curve
from the absorbances of the standards and determine the hexavalent chromium levels in the water samples by direct comparison.

A single laboratory has used this procedure to determine
the precision and recovery using tap water spiked with

hexavalent chromium at a level of 50 µg/L. The coefficient of
variation was 5.2%, and the recovery was 96%.

5.10.2. Canadian DOE Methods 24002 and 24003 for Chromium

Method 24002 is applicable to the determination of chromium
in a wide variety of waters. Samples are prepared in accord
with the procedures described in Section 5.2.2. Samples and
standards are aspirated into the air-acetylene flame, and the
absorbances are measured at a wavelength of 358.0 nm. The
chromium content of the samples is determined by direct comparison.

Method 24003 describes a chelation-extraction procedure
for the determination of low levels, 0.001 to 0.020 ppm, of
chromium in a wide variety of waters. To blanks, standards,
and aliquots of prepared samples, adjusted to pH 1.6 with
nitric acid and diluted to 100 mL with high-purity water, contained in 250-mL volumetric flasks, add 0.5 mL of bromine
water (1%) and warm on a water bath until the color of the
bromine has disappeared. Cool, and adjust the pH to 3.5 by
adding sufficient (2.0 mL) buffer (prepared by dissolving
100 g of ammonium acetate in high-purity water, adding 40 mL
of 25% v/v ammonium hydroxide and diluting to 1 L). Mix well,
add 3.0 mL of 1% ammonium pyrrolidine dithiocarbamate, and
mix again. Add 5.0 mL of methyl isobutyl ketone to each, and
shake the flasks for 5 min. Allow the layers to separate,
add high-purity water down the sides of the flasks to float
the organic phases from each standard and sample into the
necks of the flasks and aspirate into the air-acetylene flame.
Measure the absorbances at 358.0 nm. Prepare a calibration
curve from the absorbances of the standards, and determine the
chromium content of the samples by direct comparison.

In a single laboratory, the coefficient of variation at a
chromium concentration of 10 µg/L was 2.9%.

5.11. Procedures for Cobalt

Cobalt is conveniently determined in the air-acetylene
flame at a wavelength of 240.7 nm. Alternate wavelengths are
242.5 nm, 241.2 nm, 252.1 nm, and 243.6 nm. The nitrous
oxide-acetylene flame may also be used with a modest loss in

sensitivity.   There are no special difficulties associated
with electrothermal atomization techniques.   High concentrations of nitric acid however, do, show a suppressive effect.

5.11.1. U.S. EPA Methods 219.1 and 219.2 for Cobalt

Method 219.1 is applicable to the determination of cobalt
in waste water at concentrations ranging from 0.5 mg/L to 5
mg/L.   For cobalt levels below 100 µg/L, the special extraction procedure described in Section 3.2.1 may be used to pre-
treat prepared samples prior to determining cobalt by this
method.

Samples prepared by the procedures described in Section
5.2.1 and standards are aspirated into the air-acetylene flame,
and the absorbances are measured at 240.7 nm.   If the extraction procedure is used, both samples and standards must be
treated in accord with Section 3.2.1.   Prepare the calibration
curve from the absorbances of the standards and determine the
cobalt content of the samples by direct comparison.

Using a sample of mixed domestic-industrial effluent spiked
at cobalt levels of 0.20 mg/L, 1.0 mg/L, and 5.0 mg/L, a
single laboratory demonstrated precision corresponding to
coefficients of variation of 6.5%, 1%, and 1%, respectively.
The recoveries of cobalt at these levels were 98%, 98%, and
97%, respectively.

Method 219.2 describes the determination of cobalt in waste
water by furnace atomic absorption spectrometry.   The working
range of the method is from 5 µg/L to 100 µg/L.   Samples are
prepared in accord with the procedures described in Section
5.2.1.   Replicate 20-µL injections of samples and standards
are measured using the following time-temperature program:
dry, 30 s at 125°C; ash, 30 s at 900°C; atomize, 10 s at
2700°C.   The absorbance measurements are made at a wavelength
of 240.7 nm.   For every sample matrix analyzed, verification
is necessary to determine that the method of standard additions
is not required.

5.11.2. Canadian DOE Methods 27001 and 27002 for Cobalt

Method 27001 is applicable to the determination of cobalt
in a wide variety of waters.   Samples are prepared by the
procedures described in Section 5.2.2.   Samples and standards

are aspirated into the air-acetylene flame, and the absorbances are measured at 240.7 nm. The cobalt content of the
samples is determined by direct comparison.

Method 27002 describes a chelation-extraction procedure
for the determination of cobalt. Prepared samples and standards are treated by the chelation-extraction procedure described in Section 5.2.2. The organic phases are aspirated
into the air-acetylene flame and absorbance measurements are
made at a wavelength of 240.7 nm. A calibration curve is
prepared from the absorbances of the standards and the cobalt
content of the samples is determined by direct comparison.

In a single laboratory, the coefficient of variation at a
cobalt level of 10 µg/L was 1.3% using the chelation-
extraction procedure.

5.12. Procedures for Copper

The determination of copper is conveniently made at a
wavelength of 324.7 nm in the air-acetylene flame. Alternate
wavelengths are 327.4 nm, 216.5 nm, and 222.6 nm. The nitrous
oxide-acetylene flame may be used with approximately three
times poorer sensitivity. The furnace technique is reported
to suffer from possible losses resulting from premature volatilization and suppression effects from nitric acid.

5.12.1. U.S. EPA Methods 220.1 and 220.2 for Copper

Method 220.1 describes the procedures for the preparation
of samples and the atomic absorption measurements associated
with the determination of copper in waste waters. The working
range of the method is from 0.2 mg/L to 5 mg/L. Samples are
prepared in accord with Section 5.2.1. If the copper levels
are below 50 µg/L, samples and standards should be prepared by
the chelation-extraction procedure described in Section 3.2.1.
Samples and standards are aspirated into the air acetylene
flame, and the absorbances are measured at 324.7 nm. The calibration curve is prepared from the absorbances of the standards, and the copper content of the samples is determined by
direct comparison.

Some seven dozen laboratories analyzed six natural water
samples spiked with varying amounts of aluminum, cadmium,

chromium, copper, iron, manganese, lead, and zinc. For a
copper level of 315 µg/L, the coefficient of variation was 17%
and the mean value showed a 1.5% negative bias. For those
samples spiked at the 67-µg/L level, the coefficient of variation was 34% and the mean showed a 4% positive bias.

Method 220.2 describes the determination of copper in waste
water by furnace atomic absorption spectrometry. The working
range of the method is from 5 µg/L to 100 µg/L. The absorbances of replicate 20-µL aliquots of samples prepared in accord with Section 5.2.1. and of standards are measured at
324.7 nm using the following time-temperature program: dry,
30 s at 125°C; ash, 30 s at 900°C; atomize, 10 s at 2700°C.
Background correction is required if the dissolved solids are
high, and the method of standard additions is required if it
cannot be shown to be unnecessary.

5.12.2. Canadian DOE Methods 29006 and 29005 for Copper

Method 29006 is applicable to the determination of copper
in a wide variety of waters. Samples are prepared by the procedures contained in Section 5.2.2. Samples and standards
are aspirated into the air-acetylene flame, and the absorbances are measured at a wavelength of 324.7 nm. The absorbances of the standards are used to prepare the calibration
curve, and the copper content of the samples is determined by
direct comparison.

Method 29005 is a chelation-extraction procedure for the
determination of copper in water samples at levels below 100
ppb. Samples are prepared in accord with Section 5.2.2.
Standards are treated by the same extraction procedure. The
organic phases are aspirated into the air-acetylene flame,
and the absorbances are measured at 324.7 nm. The calibration
curve is prepared from the absorbances of the standards, and
the copper content of the samples is determined by direct
comparison.

In a single laboratory, the coefficient of variation at a
copper level of 10 µg/L was 1.6% using the solvent extraction
method. Recovery of copper from spiked natural waters average 90% using the solvent extraction procedure.

5.13. Procedures for Gold

Gold is conveniently determined in the air-acetylene flame

at a wavelength of 242.8 nm. An alternate wavelength is 267.6
nm. The use of the nitrous oxide-acetylene flame results in
a fourfold loss in sensitivity. The volatility of gold limits
the ashing temperature to 500°C when the furnace technique is
employed. At this low ashing temperature, matrix interferences are more common. Hence, background correction and the
method of standard additions are recommended.

5.13.1. U.S. EPA Methods 231.1 and 231.2 for Gold

Method 231.1 describes a sample preparation procedure and
the flame atomic absorption technique for the determination of
gold in waste water. The working range of the method is from
0.5 mg/L to 20 mg/L.

Transfer a 100-mL aliquot of the well-mixed sample to a
250-mL beaker, add 3 mL of concentrated nitric acid, place on
a steam bath, and evaporate to near dryness. Cool, and add
4 mL of concentrated hydrochloric acid and 2 mL of concentrated nitric acid. Cover the beaker with a watch glass and
return it to the steam bath. Heat for 30 min, uncover the
beaker, and allow the contents to evaporate to near dryness.
Cool, add 1 mL of 1:1 nitric acid, wash down the watch glass
and inside of the beaker, and filter the contents into a
100-mL volumetric flask. Rinse the beaker with distilled
water, add the rinsings to the volumetric flask, and dilute
the contents to the mark.

Aspirate the prepared sample and standards in 0.5% nitric
acid into a lean air-acetylene flame. Measure the absorbances
at 242.8 nm. Prepare the calibration curve from the absorbances of the standards and determine the gold content of the
sample by direct comparison.

Method 231.2 describes the atomic absorption determination
of gold in waste water by the furnace technique. The working
range of the method is from 5 µg/L to 100 µg/L. Samples are
prepared by the same aqua regia digestion employed for the
flame technique (method 231.1). The absorbances of replicate
20-µL aliquots of the samples and samples spiked for the standard addition method are measured at 242.8 nm. The following
time-temperature program is recommended: dry, 30 s at 125°C;
ash, 30 s at 600°C; atomize, 10 s at 2700°C. The measurements
are made with continuous flow argon in nonpyrolytic graphite
tubes. If it can be demonstrated that the method of standard

additions is unnecessary, direct comparison can be used to determine the gold content of the sample.

5.14. Procedures for Iridium

The determination of iridium is carried out at a wavelength of 264.0 nm in a rich air-acetylene flame. Alternate wavelengths are 208.9 nm, 266.5 nm, 237.3 nm, 285.0 nm, 250.3 nm, and 254.4 nm. Numerous interferences are encountered in the air-acetylene flame. In the nitrous oxide-acetylene flame there are fewer interferences, but the sensitivity is reduced by almost a factor of 10. Few problems are encountered with the determination of iridium by the furnace technique.

5.14.1. U.S. EPA Methods 235.1 and 235.2 for Iridium

Method 235.1 is applicable to the determination of iridium in waste water in the concentration range of 20 mg/L to 500 mg/L. Samples are prepared by the aqua regia digestion employed for gold (Section 5.13.1). Samples and standards in 0.5% nitric acid are aspirated into the rich air-acetylene flame, and the absorbances are measured at 264.0 nm. Prepare a calibration curve from the absorbances of the standards and determine the iridium content of the samples by direct comparison.

Method 235.2 describes the furnace technique for the determination of iridium in waste water. The working range for this method is from 0.1 mg/L to 1.5 mg/L. Samples are prepared by the aqua regia digestion used for gold (Section 5.13.1 above). Replicate 20-μL aliquots of samples and standards are injected into the furnace, and the following time-temperature program is used to measure the absorbances: dry, 30 s at 125°C; ash, 30 s at 600°C; atomize, 10 s at 2800°C. Continuous flow argon and pyrolytic graphite are recommended. The iridium content of the samples must be determined by the method of standard additions unless it can be demonstrated to be unnecessary. If this is found to be the case, direct comparison may be used.

5.15. Procedures for Iron

Iron is frequently determined at a wavelength of 248.3 nm

in a lean air-acetylene flame. Alternate wavelengths are
248.8 nm, 271.9 nm, 302.1 nm, 252.7 nm, and 372.0 nm. Reductions in sensitivity caused by nitric acid or nickel can be
controlled by using a lean (hot) air-acetylene flame. The
nitrous oxide-acetylene flame can be used, but the sensitivity
for iron is some threefold poorer. No particular problems are
encountered in the determination of iron by the furnace
technique.

5.15.1. U.S. EPA Methods 236.1 and 236.2 for Iron

Method 236.1 is applicable to the determination of iron in
water and waste water at concentrations ranging from 0.3 mg/L
to 5 mg/L. Samples are prepared by the procedures described
in Section 5.2.1. Samples and standards in the same acid matrix are aspirated into a lean air-acetylene flame, and the
absorbances are measured at 248.3 nm. The calibration curve
is prepared from the absorbances of the standards, and the
iron content of the samples is determined by direct comparison.
The chelation-extraction procedure described in Section
3.2.1 may be employed to determine low concentrations of iron.
Some six and one-half dozen laboratories analyzed natural
water samples spiked with six concentrates containing varying
amounts of aluminum, cadmium, chromium, copper, iron, manganese, lead, and zinc. For samples spiked at the 750-ppb level,
the coefficient of variation for the iron determinations was
23%, and the mean result showed a 1% negative bias. The coefficient of variation for the results with samples spiked at
400-ppb iron was 38%, and the mean for these samples was within 0.5% of the theoretical value.
Method 236.2 describes the determination of iron in water
and waste water by the furnace technique. The range of the
method is from 5 µg/L to 100 µg/L. Samples are prepared in
accord with the procedures in Section 5.2.1, and both samples
and standards are made up in 0.5% nitric acid. Unless it can
be shown to be unnecessary, the determination of iron is made
by the standard additions method. Replicate 20-µL aliquots of
the samples and of the samples spiked with iron are injected,
and the absorbances are measured at 248.3 nm using the following time-temperature program: dry, 30 s at 125°C; ash, 30 s
at 1000°C; atomize, 10 s at 2700°C. Continuous flow argon
and nonpyrolytic graphite are recommended.

5.15.2. Canadian DOE Methods 26004 and 26005 for Iron

Method 26004 describes the determination of iron in a wide
variety of waters. The procedures contained in Section 5.2.2
are used to prepare the samples, and samples and standards in
0.5% nitric acid are aspirated into the air-acetylene flame.
The absorbances are measured at 248.3 nm. The calibration
curve is prepared from the absorbances of the standards, and
the iron content of the samples is determined by direct comparison. Coefficients of variation for the determination of
iron at 0.05 mg/L, 0.1 mg/L, and 0.2 mg/L were 12%, 10%, and
5%, respectively.

The chelation-extraction procedure for the determination of
iron is described in Method 26005. Samples are prepared and
the samples and standards are extracted in accord with Section
5.2.2. The ketone layers are aspirated into the air-acetylene
flame, and the absorbances are measured at a wavelength of
248.3 nm. The iron content is determined by direct comparison
using the absorbances of the standards for the calibration
curve.

5.16. Procedures for Lead

The determination of lead is accomplished with lesser sensitivity than are the determinations of most other metals by
flame atomic absorption spectrometry. The measurements are
usually made in the air-acetylene flame at 283.3 nm. As an
alternate wavelength, 217.0 nm may be used. Both the EDL and
the HCL are available as sources of resonance radiation. Use
of the nitrous oxide-acetylene flame is possible, but a threefold loss in sensitivity results. Sensitivity is greatly improved in the graphite furnace, but the volatility of some
lead compounds limits the ashing temperature to 500°C. The
furnace technique is vulnerable to matrix interferences.
Background correction and the method of standard additions are
frequently employed to overcome these difficulties.

5.16.1. U.S. EPA Methods 239.1 and 239.2 for Lead

Method 239.1 describes the determination of lead in water
and waste water. The working range of the method is from 1
mg/L to 20 mg/L. Samples are prepared by the procedures in

Section 5.2.1. Samples and standards in 1% nitric acid are
aspirated into the air-acetylene flame, and the absorbances
are measured at 283.3 nm. The calibration curve is prepared
from the absorbances of the standards, and the lead levels of
the samples are determined by direct comparison.

The drinking water maximum for lead is 0.05 mg/L, well below the working range of method 239.1. The chelation-
extraction procedure described in Section 3.2.1 may be used in
such instances. Prepared samples and standards are extracted,
the extracts evaporated, the residues dissolved, and the resulting solutions aspirated into the air-acetylene flame. Absorbances are measured at 283.3 nm, and the lead levels of the
samples are determined by direct comparison.

In an interlaboratory study, samples of natural water
spiked with six concentrates containing varying amounts of
aluminum, cadmium, chromium, copper, iron, manganese, lead,
and zinc were analyzed by some six dozen laboratories. For
the samples spiked with 350-ppb lead, the results showed a
coefficient of variation of 34%, and the mean showed a 2% negative bias. The corresponding values for samples spiked at 90
ppb were 48% and 1% positive bias, respectively.

Method 239.2 describes the furnace technique for the determination of lead in water and waste water samples over the
concentration range of from 5 µg/L to 100 µg/L. Samples are
prepared in accord with the procedures contained in Section
5.2.1. The absorbances of replicate 20-µL injections of prepared samples and standards containing 5000-ppm lanthanum and
made up in 0.5% nitric acid are measured at a wavelength of
283.3 nm under the following conditions: dry, 30 s at 125°C;
ash, 30 s at 500°C; atomize, 10 s at 2700°C. The standard
additions method is preferred, and it must be used if it cannot be shown to be unnecessary. Background correction is
recommended.

A single laboratory using tap water spiked at lead levels
of 25 µg/L, 50 µg/L, and 100 µg/L obtained results with coefficients of variation of 5%, 3%, and 4%, respectively. Recoveries at these levels were 88%, 92%, and 95%, respectively.

5.16.2. Canadian DOE Methods 82001 and 82002 for Lead

Method 82001 is applicable to the determination of lead in
a wide variety of waters, including surface waters and

industrial waste waters. Samples are prepared by the procedures described in Section 5.2.2. Prepared samples and standards made up in 1% nitric acid are aspirated into the air-
acetylene flame, and the absorbances are measured at 283.3 nm.
The lead levels of the samples are determined by direct comparison using the absorbances of the standards for the calibration
curve.

Method 82002 describes the determination of low concentrations of lead by chelation-extraction. Samples are prepared
by, and the prepared samples and standards are extracted by,
the procedures of Section 5.2.2. The organic phases are aspirated into the air-acetylene flame, and the absorbances are
measured at 283.3 nm. The calibration curve is prepared from
the absorbances of the standards, and the lead levels of the
samples are determined by direct comparison.

In a single laboratory, the coefficient of variation at a
lead level of 10 µg/L was 2.2% using the chelation-extraction
procedure.

5.17. Procedures for Lithium

Lithium is conveniently determined in the air-acetylene
flame at a wavelength of 670.8 nm. The nitrous oxide-
acetylene flame can also be used, but the sensitivity is reduced slightly. The volatility of some lithium compounds limits the ashing temperatures for electrothermal atomization to
500°C. Hence, interferences are encountered with this
technique.

5.17.1. Canadian DOE Method 03001 for Lithium

Method 03001 is applicable to the determination of lithium
in a wide variety of waters, including surface water and industrial waste waters. Samples are prepared by the procedures
described in Section 5.2.2. The prepared samples and standards are aspirated into the air-acetylene flame, and the
absorbances are measured at 670.8 nm. The calibration curve
is prepared from the absorbances of the standards, and the
lithium content of the samples is determined by direct
comparison.

5.18. Procedures for Magnesium

The determination of magnesium is frequently made in the air-acetylene flame at a wavelength of 285.2 nm. Interferences from aluminum and silicon are overcome by the addition of a lanthanum releaser. Atomization in the nitrous oxide-acetylene flame also eliminates these interferences, but the sensitivity is some threefold poorer in this flame than in the air-acetylene flame.

5.18.1. U.S. EPA Method 242.1 for Magnesium

This method describes the determination of magnesium in waste water. The working range of the method is from 0.02 mg/L to 0.5 mg/L. Samples are prepared by the procedures of Section 5.2.1. To each 10 mL of prepared sample, add 1 mL of lanthanum solution (prepared by dissolving 29 g of lanthanum oxide in 250 mL of concentrated hydrochloric acid and diluting the resulting solution to 500 mL with high-purity water). The samples and standards, also treated with lanthanum solution, are aspirated into the air-acetylene flame, and the absorbances are measured at 285.2 nm. The magnesium content of the samples is determined by direct comparison using the absorbances of the standards for the calibration curve.

In a single laboratory using distilled water spiked at 2.1 mg/L and 8.2 mg/L, the coefficients of variation for the results of the magnesium determinations were 4% and 2%, respectively. The recoveries of magnesium, at these levels, were 100% in both cases.

5.18.2. Canadian DOE Method 12102 for Magnesium

This method is applicable to the determination of magnesium in a wide variety of waters, including surface waters and industrial waste waters. The working range of the method is from 0.01 mg/L to 2 mg/L. To 25-mL portions of the samples contained in 50 mL volumetric flasks and to measured volumes of the standards, add 0.5 mL of lanthanum solution (prepared by dissolving 117.2 g of lanthanum oxide in 200 mL of concentrated hydrochloric acid and diluting the resulting solution to 1 L with high-purity water) and bring to volume. Aspirate the samples and standards into the air-acetylene flame and measure the absorbances at 285.2 nm. Prepare the calibration curve from the absorbances of the standards and determine the magnesium content of the samples by direct comparison.

In a single laboratory, the coefficient of variation at a magnesium level of 2 mg/L was 0.5%.

5.19. Procedures for Manganese

Manganese is frequently determined at a wavelength of 279.5 nm in a lean air-acetylene flame. Alternate wavelengths are 279.8 nm and 280.1 nm. Manganese may also be determined in the nitrous oxide-acetylene flame, but there is a three-fold loss in sensitivity. The determination of manganese by the furnace technique presents no unusual difficulties.

5.19.1. U.S. EPA Methods 243.1 and 243.2 for Manganese

Method 243.1 describes the determination of manganese in waste water. The working range of the method is from 0.1 mg/L to 3 mg/L. Lower manganese concentrations may be determined by the chelation-extraction procedure in Section 3.2.1. Wastewater samples are prepared in accord with Section 5.2.1. Samples and standards, matched in terms of matrix acid and preconcentrated by chelation-extraction if necessary, are aspirated into the air-acetylene flame. The absorbances are measured at 279.5 nm. Using the absorbances of the standards to prepare the calibration curve, the manganese contents of the samples are determined by direct comparison.

Some six dozen laboratories analyzed natural water samples spiked with six concentrates containing varying amounts of aluminum, cadmium, chromium, copper, iron, manganese, lead, and zinc. For the samples spiked with 450-ppb manganese, the coefficient of variation for the results was 19%, and the mean showed a positive bias of 1.8%. The corresponding values for samples spiked at the 95-ppb level were 30% and 0%, respectively.

Method 243.2 describes the determination of manganese in waste water by the furnace technique. The optimum concentration for this method is from 1 μg/L to 30 μg/L. Samples are prepared by the procedure in Section 5.2.1, and the absorbances of replicate 20-μL injections of samples and standards are measured at 279.5 nm using the following time-temperature program: dry, 30 s at 125°C; ash, 30 s at 1000°C; atomize, 10 s at 2700°C. Continuous flow argon and nonpyrolytic graphite are recommended. Unless it can be shown to be unnecessary, the method of standard additions must be used to determine the manganese content of the samples.

5.19.2. Canadian DOE Methods 25004 and 25005 for Manganese

Method 25004 describes the determination of manganese in a wide variety of waters. Samples are prepared by the procedures in Section 5.2.2, and the samples and standards are aspirated into the air-acetylene flame. The absorbances are measured at 279.8 nm, and the manganese content of the samples is determined by direct comparison using the absorbances of the standards to prepare the calibration curve.

Method 25005 describes the chelation-extraction procedure for the determination of low levels (1 μg/L to 20 μg/L) of manganese in a wide variety of waters. Samples are prepared by the procedures in Section 5.2.2, and samples and standards are treated as follows.

Adjust the pH of 100 mL aliquots of the samples and standards contained in 250-mL volumetric flasks to the 9 to 11 range with 25% (v/v) ammonium hydroxide. Add 6 mL of 1% 8-hydroxyquinoline in MIBK to each flask and shake the flasks for 5 min. Allow the phases to separate, and float the organic layer in each flask up into the necks by adding high-purity water down the sides of the flasks. Aspirate the organic layers into the air-acetylene flame and measure their absorbances at 279.8 nm. Prepare the calibration curve from the absorbances of the standards and determine the manganese contents of the samples by direct comparison.

In a single laboratory, the coefficient of variation at a manganese level of 10 μg/L was 4% using method 25005.

5.20. Procedures for Mercury

Mercury is conveniently determined by the cold vapor technique at a wavelength of 253.7 nm. The volatility of mercury demands special handling and preparation procedures to avoid losses.

5.20.1. U.S. EPA Methods 245.1 and 245.2 for Mercury

Method 245.1 is applicable to the determination of mercury in drinking water, surface water, saline water, and waste water. The working range of the method is from 0.5 µg/L to 10 µg/L. Samples and standards are prepared as follows.

Transfer 100-mL aliquots of well-mixed samples and appropriate volumes of mercury standard diluted to 100 mL to separate biochemical oxygen demand (BOD) bottles of 300-mL capacity. To each bottle, add 5 mL of 0.5 N sulfuric acid, 2.5 mL of concentrated nitric acid, and 15 mL of 5% (w/v) potassium permanganate solution. Mix well and allow the bottles to stand for 15 min. If the purple color is discharged during this time, add additional 15-mL portions of the potassium permanganate solution until a purple color persists for 15 min. Add 8 mL of 5% (w/v) potassium persulfate solution to each bottle and heat them in a 95°C water bath for 2 h.

The absorbances of the samples and standards are measured individually from this point on. Cool, and add 6 mL of sodium chloride-hydroxylamine sulfate* solution prepared by dissolving 12 g of each in water and diluting the resulting solution to 100 mL. Quickly inject 5 mL of 10% (w/v) stannous sulfate* suspension in 0.5 N sulfuric acid* into the BOD bottle, and immediately attach the bottle to the aeration apparatus which is connected to the absorption cell in the optical path of the atomic absorption spectrometer. When the absorbance reading becomes stable, open the bypass and continue aeration until the absorbance reading returns to its baseline. Remove the BOD bottle and proceed with subsequent measurements.

Prepare a calibration curve from the maximum absorbances of the standards. The mercury content of the samples is determined by direct comparison.

A single laboratory analyzed river water samples spiked with mercury at levels of 1.0, 3.0, and 4.0 µg/L. The coefficients of variation for these measurements were 14%, 3%, and 2%, respectively, and the mercury recoveries were 89%, 87%, and 87%, respectively.

In an interlaboratory study, eight different samples of natural water, spiked with inorganic and organic mercury, were

---

*The corresponding chlorides may be used.

Table 5.1. Results of Interlaboratory Mercury Analyses

Number of participants	Mean (µg/L)	Coefficient of variation (%)	True value (µg/L)	Bias (%)
76	0.349	80	0.21	66
80	0.414	70	0.27	53
82	0.674	80	0.51	32
77	0.709	55	0.60	18
82	3.41	43	3.4	0.34
79	3.81	29	4.1	-7.1
79	8.77	42	8.8	-0.4
78	9.10	39	9.6	-5.2

evaluated. These results are summarized in Table 5.1.

The U.S. EPA has also described an automated cold vapor technique for the determination of mercury in water, method 245.2. This method may also be applicable to saline waters, waste waters, effluents, and domestic sewages. The manifold for this method is shown in Figure 5.2. The coefficients of variation for standards containing 0.5, 1.0, 2.0, 5.0, 10.0, and 20.0 µg/L were 8%, 7%, 5%, 4%, 4%, and 4%, respectively. In a single laboratory, using surface water samples spiked with 10 organic mercurials at the 10-µg/L level, recoveries ranged from 87% to 117%.

5.20.2. Canadian DOE Method 80311 for Mercury

Method 80311 describes an automated cold vapor technique for the determination of mercury in surface and ground waters as well as in rain and melted snow. The analytical range of the method is from 0.5 µg/L to 100 µg/L.

Organomercury compounds in the sample are oxidized to inorganic mercury(II) ions by heating with sulfuric acid, potassium permanganate, and potassium persulfate. The mercury(II) is then reduced to elemental mercury with hydroxylamine and stannous sulfate. The elemental mercury is then sparged from the solution with a stream of air and transported to the absorption cell in the optical path of the atomic absorption spectrometer. In addition to offering the advantages of speed

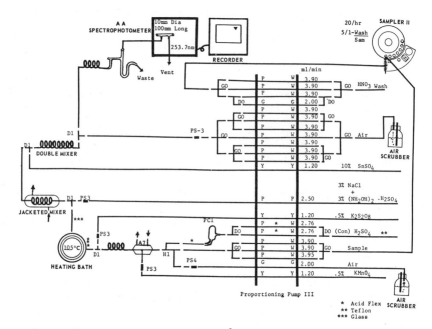

Figure 5.2. Mercury manifold.[9] P = purple; Y = yellow;
G = green; W = white.

and reliability, the automated procedure provides a closed
system in which the determination can take place without
losses of volatile mercury compounds.

In a single laboratory, the coefficient of variation at
mercury concentrations of 0.07 µg/L, 0.28 µg/L, and 0.55 µg/L
were 6%, 4%, and 1%, respectively. Recoveries of inorganic
and organic mercury averaged 95%.

5.21. Procedures for Molybdenum

Sensitivity for the determination of molybdenum is, in general, poorer than that with which most other metals are determined. Absorbance measurements are made at a wavelength of
313.3 nm in a rich nitrous oxide-acetylene flame. Alternate
wavelengths are 317.0 nm, 379.8 nm, 319.4 nm, 386.4 nm,
390.3 nm, and 315.8 nm. Sensitivity is poorer by a factor of
two in the air-acetylene flame. Calcium interference can be
controlled by the addition of aluminum or other refractory

metal. The furnace technique also suffers from interferences.
Hence, the method of standard additions is frequently used.

5.21.1. U.S. EPA Methods 246.1 and 246.2 for Molybdenum

Method 246.1 describes the determination of molybdenum in
waste water by flame atomic absorption spectrometry over the
concentration range of 1 mg/L to 40 mg/L. Samples are prepared by the methods in Section 5.2.1, and each 100 mL of
prepared sample and standard is treated with 2 mL of aluminum
nitrate solution prepared by dissolving 139 g of the nonahy-
drated salt in 150 mL of water with heating and diluting the
resulting solution to 200 mL after cooling to room temperature. The treated samples and standards are then aspirated
into a rich nitrous oxide-acetylene flame, and the absorbances
are measured at 313.3 nm. The absorbances of the standards
are used to prepare the calibration curve, and the molybdenum
contents of the samples are determined by direct comparison.

In a single laboratory, using mixed industrial-domestic
effluent spiked to molybdenum levels of 0.30 mg/L, 1.5 mg/L,
and 7.5 mg/L, the coefficients of variation were 2%, 1%, and
1%, respectively. The recoveries were 100%, 96%, and 95%,
respectively.

The furnace technique for the determination of molybdenum
is described in method 246.2. The working range of this method is from 3 µg/L to 60 µg/L. Sample preparation is in accord
with the procedures of Section 5.2.1. The determination of
molybdenum is made by the method of standard additions unless
it has been shown to be unnecessary. The samples and spiked
samples should be in 0.5% nitric acid. The absorbances of
replicate 20-µL injections are measured at 313.3 nm using the
following time-temperature program: dry, 30 s at 125°C; ash,
30 s at 1400°C; atomize, 15 s at 2800°C.

5.21.2. Canadian DOE Methods 42001 and 42002 for Molybdenum

Method 42001 is applicable to the determination of molybdenum in a wide variety of waters, including surface waters
and industrial waste waters. Samples are prepared by the procedures described in Section 5.2.2, and the prepared samples
and standards are aspirated into the nitrous oxide-acetylene
flame. Absorbance measurements are made at 313.5 nm. The

calibration curve is prepared from the absorbances of the
standards, and the molybdenum content of the samples is determined by direct comparison.

Method 42002 describes a chelation-extraction procedure for
the determination of low levels, 0.2 µg/L to 10 µg/L, of molybdenum in water and waste water. The samples are prepared by
the procedures in Section 5.2.2, and the prepared samples and
standards are treated as follows: Pipet 150 mL aliquots of
the prepared samples and measured volumes of molybdenum standard diluted to 150 mL into 200-mL volumetric flasks. Adjust
the contents of each flask to pH 1.6, add 0.5 mL of 1% bromine
water, and heat on a water bath until the color of bromine
disappears. Cool, and add 5 mL of 1% benzoin α-oxime in ethanol. Add 3.0 mL of n-butyl acetate to each, and shake the
flasks for 5 min. Allow the layers to separate and float the
organic layers up into the necks of the flasks by slowly adding high-purity water down the sides. Aspirate the organic
layers into the nitrous oxide-acetylene flame and measure the
absorbances at 313.5 nm. Determine the molybdenum content of
the samples by direct comparison using the absorbances of the
standards for the calibration curve.

5.22. Procedures for Nickel

Nickel is most frequently determined in the air-acetylene
flame at 232.0 nm using a spectral band pass no larger than
0.2 nm. Parasitic radiation from the emissions at 231.7 nm
and 232.1 nm can cause curvature of the calibration plot and
loss of sensitivity. Alternate wavelengths are 231.1 nm,
352.5 nm, 341.5 nm, and 305.1 nm. Use of the nitrous oxide-
acetylene flame results in a threefold loss in sensitivity.
The low volatility of most nickel compounds allows high ashing
temperatures to be used in the furnace program. While matrix
interferences can be eliminated in this way, spectral interferences remain. Hence, the standard additions method is frequently used for the determination of nickel.

5.22.1. U.S. EPA Methods 249.1 and 249.2 for Nickel

Method 249.1 is applicable to the determination of nickel
in waste water over the concentration range of 0.3 mg/L to

5 mg/L. Samples are prepared in accord with the procedures
described in Section 5.2.1. The chelation-extraction procedure described in Section 3.2.1 is recommended for nickel
levels below 0.1 mg/L. The prepared and possibly extracted
samples and standards are aspirated into the air-acetylene
flame, and the absorbances are measured at 232.0 nm using a
spectral band pass no greater than 0.2 nm. The calibration
curve is prepared from the absorbances of the standards, and
the nickel content of the samples is determined by direct
comparison.

In a single laboratory, using mixed industrial-domestic
effluent spiked at nickel levels of 0.2 mg/L, 1 mg/L, and
5 mg/L, the coefficients of variation were 6%, 2%, and 0.8%,
respectively. The corresponding recoveries were 100%, 97%,
and 93%.

Method 249.2 describes the furnace technique for the determination of nickel in waste water in the concentration range
of from 5 µg/L to 100 µg/L. Samples are prepared in accord
with the procedures of Section 5.2.1. It is necessary to verify for every sample matrix encountered that the method of
standard additions is unnecessary in order to determine nickel
by direct comparison. The absorbances of 20-µL injections of
samples and spiked samples are measured at 232.0 nm using the
following time-temperature program: dry, 30 s at 125°C; ash,
30 s at 900°C; atomize, 10 s at 2700°C.

5.22.2. Canadian DOE Methods 28001 and 28002 for Nickel

Method 28001 is applicable to the determination of nickel
in a wide variety of waters and waste waters. The samples
are prepared by the procedures described in Section 5.2.2, and
the samples and standards are aspirated into the air-acetylene
flame. Absorbances are measured at 232.0 nm, and the nickel
content of the samples is determined by direct comparison
using the absorbances of the standards to prepare the calibration curve.

Method 28002 is the chelation-extraction procedure for the
determination of low levels of nickel. Samples are prepared
by, and samples and standards are extracted by, the procedures
described in Section 5.2.2. The organic phases are aspirated
into the air-acetylene flame, and the absorbances are measured
at 232.0 nm. The calibration curve is prepared from the

absorbances of the standards, and the nickel content of the samples is determined by direct comparison.

In a single laboratory, the coefficient of variation at a nickel level of 10 µg/L was 3.8% using the chelation-extraction procedure.

5.23. Procedures for Osmium

The sensitivity for the determination of osmium by atomic absorption spectrometry is poorer than that for most other metals. Absorbance measurements are made at 290.9 nm in a rich nitrous oxide-acetylene flame. Alternate wavelengths are 305.9 nm, 263.7 nm, 301.8 nm, and 330.2 nm. Sensitivity is five fold poorer in the air-acetylene flame.

5.23.1. U.S. EPA Methods 252.1 and 252.2 for Osmium

Method 252.1 describes the determination of osmium in waste water over the concentration range of 2 mg/L to 100 mg/L. Samples are prepared as follows.

Transfer a 100-mL aliquot of the well-mixed sample to a 250-mL beaker, add 1 mL of concentrated nitric acid, and warm on a hotplate for 15 min. Cool, and filter into a 100-mL volumetric flask. Add 1 mL of concentrated sulfuric acid and -bring to volume. Aspirate samples and standards into the rich nitrous oxide-acetylene flame, and measure the absorbances at 290.9 nm. The calibration curve is prepared from the absorbances of the standards, and the osmium content of the samples is determined by direct comparison.

Method 252.2 describes the furnace technique for the determination of low levels, 50 µg/L to 500 µg/L, of osmium in waste water. Samples are prepared as follows.

Transfer a 100-mL aliquot of the well-mixed sample to a 250-mL beaker, add 1 mL of concentrated nitric acid, and warm on a hotplate for 15 min. Cool, filter into a 100-mL volumetric flask, and bring to volume with high-purity water.

For each sample matrix encountered, it is necessary to establish the optimum ashing time and temperature, and it is necessary to demonstrate that the method of standard additions is unnecessary in order to determine osmium by direct comparison. The absorbances of 20-µL injections of samples and spiked samples are measured at 209.9 nm. A 30-s drying stage at 105°C, and a 10-s atomization stage at 2700°C are recommended. Background correction, continuous flow argon, and nonpyrolytic graphite are also recommended.

5.24. Procedures for Palladium

Palladium is frequently determined in a lean air-acetylene flame at 247.6 nm. Alternate wavelengths are 244.8 nm, 276.3 nm, and 240.5 nm. Palladium may also be determined in the nitrous oxide-acetylene flame, but there is a fivefold loss in sensitivity.

5.24.1. U.S. EPA Methods 253.1 and 253.2 for Palladium

Method 253.1 describes the determination of palladium in waste waters in the concentration range of 0.5 mg/L to 15 mg/L. The samples are prepared as follows.

Transfer a 100-mL aliquot of the well-mixed sample to a 250-mL beaker, add 3 mL of concentrated nitric acid, and evaporate to near dryness on a steam bath. Cool, and add 3 mL of concentrated hydrochloric acid and 2 mL of concentrated nitric acid. Cover with a watch glass, return to the steam bath, and continue heating for 30 min. Uncover and evaporate to near dryness. Cool, wash down the watch glass and inside walls of the beaker, and filter the contents into a volumetric flask of 25-mL, 50-mL, or 100-mL capacity. Add sufficient nitric acid so that the final solution will be 0.5% and bring to volume with high-purity water. Aspirate the prepared samples and standards into the air-acetylene flame and measure the absorbances at 247.6 nm. Prepare the calibration curve from the absorbances of the standards and determine the palladium content of the samples by direct comparison.

Method 253.2 describes the furnace technique for the determination of palladium in waste water over the concentration range of 20 μg/L to 400 μg/L. Samples are prepared by the same aqua regia digestion described above (method 253.1). The method of standard additions must be used unless it is shown to be unnecessary. If this is the case, direct comparison may be used to determine the palladium content of the samples. The absorbances of replicate 20-μL injections of samples and

spiked samples are measured at 247.6 nm using the following
time-temperature program: dry, 30 s at 125°C; ash, 30 s at
1000°C; atomize, 10 s at 2800°C. Background correction, continuous purge argon, and pyrolytic graphite are recommended.

5.25. Procedures for Platinum

The determination of platinum is made at a wavelength of
265.9 nm in the air-acetylene flame. There are interferences
by other nobel metals, but these are overcome by the addition
of lanthanum chloride or by the use of the nitrous oxide-
acetylene flame. Sensitivity is some fivefold poorer in this
flame than in the air-acetylene flame. Alternate wavelengths
are 306.5 nm, 283.0 nm, 293.0 nm, 273.4 nm, 270.2 nm, 248.7 nm,
299.8 nm, and 271.9 nm.

5.25.1. U.S. EPA Methods 255.1 and 255.2 for Platinum

Method 255.1 is applicable to the determination of platinum in waste water in the concentration range of 5 mg/L to
75 mg/L. The samples are prepared by the same aqua regia
digestion used for palladium in Section 5.24.1. Prepared
samples and standards are aspirated into the air-acetylene
flame, and the absorbances are measured at 265.9 nm. The
platinum content of the samples is determined by direct comparison using the absorbances of the standards for the calibration curve.

Method 255.2 describes the furnace technique for the determination of platinum in waste water at concentrations of
0.1 mg/L to 2 mg/L. The aqua regia digestion in Section
5.24.1 is used to prepare the samples. The method of standard
additions must be used to determine the platinum content of
the samples unless it is shown to be unnecessary. In such
cases, direct comparison may be used. The absorbances of
replicate 20-µL injections of the prepared samples and the
standards or spiked samples are measured at 265.9 nm using the
following time-temperature program: dry, 30 s at 125°C; ash,
30 s at 1300°C; atomize, 10 s at 2800°C. Background correction, continuous flow argon, and pyrolytic graphite are
recommended.

5.26. Procedures for Potassium

Potassium is most frequently determined at 766.5 nm in the air-acetylene flame. Under these conditions, there is ionization interference. Hence, both the samples and the standards are made up in 0.1% sodium. An alternate wavelength is 769.9 nm. The photocathode material is not highly responsive to these wavelengths. Some instruments, therefore, may show poor sensitivity for potassium.

5.26.1. U.S. EPA Method 258.1 for Potassium

Method 258.1 is applicable to the determination of potassium in waste water over the 0.1 mg/L to 2 mg/L concentration range. The samples are prepared by the procedures described in Section 5.2.1. To each 100 mL of prepared sample and standard, 10 mL of 3% sodium chloride solution are added, and the absorbances are measured at 766.5 nm in the air-acetylene flame. The potassium content of the samples is determined by direct comparison using the absorbances of the standards for the calibration curve.

In a single laboratory, using distilled water spiked at potassium levels of 1.6 mg/L and 6.3 mg/L, the coefficients of variation were 13% and 8%, respectively. The recoveries were 103% and 102%, respectively.

5.27. Procedures for Rhenium

Rhenium is determined at a wavelength of 346.0 nm in a rich nitrous oxide-acetylene flame. Alternate wavelengths are 346.5 nm and 345.2 nm. The sensitivity of atomic absorption spectrometry for rhenium is far poorer than it is for the determination of most other metals.

5.27.1. U.S. EPA Methods 264.1 and 264.2 for Rhenium

Method 264.1 describes the sample preparation and flame atomic absorption technique for the determination of rhenium. The working range of the method is from 50 mg/L to 1000 mg/L. The samples are prepared by procedures identical to those used

for osmium in Section 5.23.1, method 252.2. The prepared
samples and standards are aspirated into the nitrous oxide-
acetylene flame, and the absorbances at 346.0 nm are recorded.
The calibration curve is prepared from the absorbances of the
standards, and the rhenium content of the samples is determined
by direct comparison.

Method 264.2 describes the furnace technique for the determination of rhenium. The working range of the method is from
0.5 mg/L to 5 mg/L. The samples are prepared by the procedures cited above, Section 5.23.1, method 252.2. For every
sample matrix analyzed, verification is necessary to determine that the method of standard additions is not required.
Since many rhenium compounds volatilize near 300°C, the ashing
temperature should also be verified for each sample matrix
analyzed. The absorbances of replicate 20-μL injections of
prepared samples and standards are measured at 346.0 nm using
the following recommendations for the time-temperature program: dry, 30 s at 125°C; ash, 30 s at 300°C; atomize, 10 s
at 2800°C. Background correction is required for samples
having a high total dissolved solids content. Continuous flow
argon and pyrolytic graphite are recommended.

5.28. Procedures for Rhodium

Rhodium can be determined in the air-acetylene flame at a
wavelength of 343.5 nm. Alternate wavelengths are 369.2 nm,
339.7 nm, 350.2 nm, 365.8 nm, and 370.1 nm. There are numerous interferences, most of which are eliminated in the nitrous
oxide-acetylene flame. Use of this flame, however, results in
a threefold loss of sensitivity relative to the air-acetylene
flame.

5.28.1. U.S. EPA Methods 265.1 and 265.2 for Rhodium

Method 265.1 is applicable to the determination of rhodium
in waste water at concentrations ranging from 1 mg/L to 30 mg/
L. The samples are prepared by the same aqua regia digestion
used to prepare samples for palladium analyses, Section 5.24.1.
Prepared samples and standards are aspirated into the air-
acetylene flame, and the absorbances at 343.5 nm are recorded.
The calibration curve is prepared from the absorbances of the
standards, and the rhodium content of the samples is determined by direct comparison.

Method 265.2 describes the furnace technique for the determination of low levels, 20 µg/L to 400 µg/L, of rhodium in waste water. Samples are prepared by the aqua regia digestion described in Section 5.24.1. It is necessary to demonstrate that the method of standard additions is not needed in order to employ direct comparison for the determination of rhodium by the furnace technique. The absorbances of replicate 20-µL injections of prepared samples and standards are measured at 343.5 nm using the following time-temperature program: dry, 30 s at 125°C; ash, 30 s at 1200°C; atomize, 10 s at 2800°C. Background correction is required for samples having high total dissolved solids levels. Continuous flow argon and pyrolytic graphite are recommended.

5.29. Procedures for Ruthenium

Ruthenium can be determined in either the air-acetylene flame or in the nitrous oxide-acetylene flame at 349.9 nm. The former flame shows a sixfold better sensitivity but more interferences than the latter. The addition of lanthanum chloride eliminates the interferences and improves the sensitivity. Alternate wavelengths are 372.8 nm and 379.9 nm.

5.29.1. U.S. EPA Methods 267.1 and 267.2 for Ruthenium

Method 267.1 describes the sample preparation and flame atomic absorption techniques for the determination of from 1 mg/L to 50 mg/L of ruthenium in waste waters. Samples are prepared as follows.

Transfer a 100-mL aliquot of the well-mixed sample to a 250-mL beaker, add 2 mL of 1:1 redistilled hydrochloric acid-high-purity water, and warm on a 95°C steam bath for 15 min. Cool, and filter into a 100-mL volumetric flask. Bring to volume with high-purity water. Aspirate the prepared sample and the standards into the air-acetylene flame and measure the absorbances at 349.9 nm. Determine the ruthenium content of the samples by direct comparison using the absorbances of the standards to establish the calibration curve.

Method 267.2 describes the furnace technique for the determination of ruthenium in the concentration range of 0.1 mg/L to 2 mg/L. Samples are prepared by the procedure described in method 267.1. In order to determine ruthenium by direct comparison, it is necessary to demonstrate that the method of standard additions is not needed. The absorbances of replicate 20-μL injections of prepared samples and standards are measured at 349.9 nm. The following time-temperature program is recommended: dry, 30 s at 125°C; ash, 30 s at 400°C; atomize, 10 s at 2800°C. Background corrections are required when the absorbances of samples with high total dissolved solids are measured. Continuous flow argon and non-pyrolytic graphite are recommended.

5.30. Procedures for Selenium

The most commonly encountered procedures for the determination of selenium involve generation of the gaseous hydride, hydrogen selenide, and its subsequent thermal decomposition and atomization. Absorbance measurements are made at 196.0 nm. Alternate wavelengths are 204.0 nm, 206.3 nm, and 207.5 nm. The argon-hydrogen flame is recommended for decomposition-atomization because of its transparency to short-wavelength radiation, and the EDL is preferred to the HCL as a source of resonance radiation because of the more intense output of the former. It is also possible to carry out the decomposition-atomization of the hydride in a quartz tube heated by either a resistance wire coil or a flame. The composition of the flame used to heat the quartz tube is of little consequence in the absorbance measurements. Although the graphite furnace technique is some 100 times more sensitive than the gaseous hydride technique, the former suffers more interferences. The volatility of selenium compounds limits the ashing temperature. Hence, matrix interferences are common. The addition of 1% nickel nitrate minimizes the interfering effects of chloride and sulfate.

5.30.1. U.S. EPA Methods 270.2 and 270.3 for Selenium

Method 270.2 describes the furnace technique for the determination of selenium in the concentration range of 5 μg/L to 100 μg/L in both drinking water samples and waste water samples. The samples are prepared for furnace atomic absorption spectrometry as follows.

Transfer a 100-mL aliquot of the well-mixed sample to a

250-mL beaker; add 1 mL of concentrated nitric acid and 2 mL
of 30% hydrogen peroxide. Heat to just below boiling for 1 h
or until the volume is reduced by more than half. Cool, and
dilute to 50 mL with high-purity water. Pipette a 5-mL aliquot of this solution into a 10-mL volumetric flask, add 1 mL
of 1% nickel nitrate solution (made by dissolving 24.78 g of
the hexahydrated salt in water and diluting the resulting solution to 500 mL), and dilute to 10 mL. Replicate 20-μL aliquots from the sample in the 10-mL flask and from the standards made up with appropriate amounts of nitric acid, hydrogen peroxide, and nickel nitrate are injected into the graphite furnace. The absorbances are measured at 196.0 nm. The
following time-temperature program is recommended: dry, 30 s
at 125°C; ash, 30 s at 1200°C; atomize, 10 s at 2700°C. The
method of standard additions must be used to determine the
selenium content of the samples unless it has been demonstrated
that direct comparison is free of interferences. Background
correction, continuous flow argon, and nonpyrolitic graphite
are recommended.

Using the effluent from a sewage treatment plant spiked to
a selenium level of 20 μg/L, a recovery of 99% was obtained.
With industrial effluent spiked to 50 μg/L, recoveries ranged
from 94% to 112%. In a 0.1% nickel nitrate matrix with selenium concentrations of 5 μg/L, 10 μg/L, 20 μg/L, 40 μg/L,
50 μg/L, and 100 μg/L, the coefficients of variation were
14.2%, 11.6%, 9.3%, 7.2%, 6.4%, and 4.1%, respectively. In
a single laboratory, using tap water spiked with selenium at
concentrations of 5 μg/L, 10 μg/L, and 20 μg/L, the coefficients of variation were 12%, 4%, and 2.5%. respectively. The
corresponding recoveries were 92%, 98%, and 100%.

Method 270.3 describes the hydride generation technique for
the determination of selenium in drinking water, surface and
ground water, saline water, and waste water. The working
range of the method is from 2 μg/L to 20 μg/L.

Samples are prepared for the determination of selenium by
the same procedures used to prepare samples for arsenic determinations, Section 5.5.1, method 206.5. The generation of
hydrogen selenide and its subsequent atomization in the argon-
hydrogen flame are carried out by essentially the same techniques as those described in Section 5.5.1, method 206.3, for
arsenic. The absorbances at 196.0 nm are recorded on a strip
chart, and the selenium content of the samples is determined

by direct comparison. Samples of industrial waste waters
should be spiked with known amounts of selenium and carried
through the procedure to establish recovery factors.

A single laboratory determined selenium in replicate samples from solutions containing 5 µg/L, 10 µg/L, and 15 µg/L.
The coefficients of variation were 12%, 11%, and 19%, respectively. The corresponding recoveries were 100%, 100%, and
101%.

5.30.2. Canadian DOE Method 34102 for Selenium

Method 34102 is applicable to the determination of selenium
in surface waters, ground waters, saline waters, and waste
waters at concentrations ranging from 0.1 µg/L to 50 µg/L.
The sample preparation and automated hydride generation-
atomic absorption system are identical to those described for
the determination of arsenic in Section 5.2.2. The same
manifold, Figure 5.1, is used. Absorbance measurements, however, are made at 196.0 nm for the determination of selenium.

In a single laboratory, the coefficients of variation at
selenium levels of 0.36 µg/L, 0.78 µg/L, 1.3 µg/L, and
3.3 µg/L were 8.9%, 3.7%, 4.2%, and 3.0%, respectively. The
average recoveries from samples spiked at selenium levels
ranging from 0.1 µg/L to 3 µg/L were from 111% to 95%.

5.31. Procedures for Silver

Silver is conveniently determined in the air-acetylene
flame at a wavelength of 328.1 nm. An alternate wavelength
of 338.3 nm may be used with only half the sensitivity. Use
of the nitrous oxide-acetylene flame results in a threefold
loss in sensitivity. The volatility of silver compounds
limits the ashing temperature to 500°C. Hence, the furnace
technique is subject to matrix interferences.

5.31.1. U.S. EPA Methods 272.1 and 272.2 for Silver

Method 272.1 is applicable to the determination of silver
in drinking water and in waste water at concentrations
ranging from 0.1 mg/L to 4 mg/L. This method makes use of
cyanogen iodide solution (prepared by dissolving 4.0 mL of

concentrated ammonia, 6.5 g of potassium cyanide, and 5.0 mL of 1.0 N iodine in 50 mL of high-purity water and diluting the resulting solution to 100 mL) to recover silver absorbed on to the walls of the sample container and/or precipitated as silver chloride. The cyanogen iodide is used at a rate of 1 mL per 100 mL of sample after the sample has been rendered ammoniacal, and the treated sample is allowed to stand for 1 h before proceeding with the analysis. The samples are prepared by the procedures described in Section 5.2.1, making certain, however, to avoid the use of hydrochloric acid. For low levels of silver, the chelation-extraction procedure in Section 3.2.1 may be employed. The prepared samples and standards carried through the chelation-extraction procedure, if it has been used, are aspirated into the air-acetylene flame, and the absorbances are measured at 328.1 nm. The silver content of the samples is determined by direct comparison using the absorbances of the silver standards for the calibration curve.

A synthetic sample containing 50 µg of silver was analyzed by 50 laboratories in a round robin. The coefficient of variation for the results was 17%, and the grand mean differed from the theoretical value by 11%.

Method 272.2 describes the furnace technique for the determination of silver in drinking water and waste water in the range of from 1 µg/L to 25 µg/L. Absorbed and/or precipitated silver is recovered with cyanogen iodide as described in method 272.1, and the samples are prepared in accord with the procedures in Section 5.2.1. The use of hydrochloric acid, however, is avoided. It is necessary to demonstrate that the method of standard additions is not needed before direct comparison may be used to determine the silver content of the samples. The absorbances of replicate 20-µL injections of the samples and the spiked samples are measured at 328.1 nm using the following time-temperature program: dry, 30 s at 125°C; ash, 30 s at 400°C; atomize, 10 s at 2700°C. Background corrections, continuous flow argon, and nonpyrolytic graphite are recommended.

In a single laboratory, using tap water spiked at silver levels of 25 µg/L, 50 µg/L, and 75 µg/L, the coefficients of variation were 1.6%, 1.4%, and 1.2%, respectively. The silver recoveries from these samples were 94%, 100%, and 104%, respectively.

5.31.2. Canadian DOE Methods 47001 and 47002 for Silver

Method 47001 is applicable to the determination of silver in a wide variety of waters. The special preservation and storage conditions in Section 5.1 should be noted, and the samples are prepared by the procedures described in Section 5.2.2. The prepared samples and standards are aspirated into the air-acetylene flame, and the absorbances are measured at 328.1 nm. Using the absorbances of the standards for the calibration curve, the silver content of the samples is determined by direct comparison.

The chelation-extraction procedure for low levels, 5 µg/L and 50 µg/L, of silver is described in method 47002. The procedure is as follows: pipette 100-mL aliquots of the samples and measured volumes of the standard diluted to 100 mL into 250-mL separatory funnels, adjust to the pH range of 3.5 to 6.5, and add 10.0 mL of 0.05% (w/v) diphenyl-thiocarbazone in ethyl propionate. Shake the funnels for 1 min using short, rapid strokes. Allow the funnels to stand for 10 min while the phases separate, drain the aqueous phases to waste, and, being certain to exclude moisture, transfer the organic phases to clean, dry test tubes with inert stoppers. Add 1 mL of acetone to the contents of each test tube to solubilize any minute amounts of water that may have been carried over into the organic phases. Aspirate the organic phases into the air-acetylene flame and measure the absorbances at 328.1 nm. Prepare the calibration curve from the absorbances of the standards and determine the silver content of the samples by direct comparison.

5.32. Procedures for Sodium

Sodium is conveniently determined in the air-acetylene flame at 589.0 nm. The effects of ionization interference can be minimized by adding excess potassium to both samples and standards. Ionization interference effects can also be eliminated by using cooler flames, air-propane or air-hydrogen.

5.32.1. U.S. EPA Method 273.1 for Sodium

Method 273.1 is applicable to the determination of sodium

in waste water at concentrations ranging from 0.03 mg/L to
1 mg/L. Samples are prepared by the procedures described in
Section 5.2.1. The prepared samples and the standards are
treated with 2% (w/v) potassium chloride solution at a ratio
of 100 mL of sample or standard to 1 mL of potassium chloride
solution. The treated samples and standards are then aspirated into the air-acetylene flame, and the absorbances are
measured at 589.6 nm. The sodium content of the samples is
determined by direct comparison using the absorbances of the
standards for the calibration curve.

In a single laboratory using distilled water spiked to
sodium levels of 8.2 mg/L and 52 mg/L, the coefficients of
variation were 1.2% and 1.3%, respectively. The corresponding
recoveries were 102% and 100%.

5.33. Procedures for Strontium

Strontium can be determined in the air-acetylene flame at a
wavelength of 460.7 nm. Chemical interferences from silicate
and phosphate can be controlled by the addition of lanthanum.
The use of the nitrous oxide-acetylene flame gives slightly
better sensitivity and freedom from chemical interferences,
but it suffers from ionization interference. The effects of
the latter can be minimized by the addition of excess sodium
or potassium. No particular difficulties are encountered
with the furnace technique.

5.33.1. Canadian DOE Method 38001 for Strontium

This method describes the determination of strontium in
water and waste water by standard additions. Samples are
prepared by the procedures in Section 5.2.2, and three 20-mL
aliquots of each are pipetted into separate Erlenmeyer flasks.
The first aliquots are treated with 5.0 mL of high-purity
water; the second aliquots are treated with 5.0 mL of a
standard estimated to be three times more concentrated in
terms of strontium than the sample, and the third aliquots are
treated with 5.0 mL of standard estimated to be five times more
concentrated in terms of strontium than the sample. The treated samples are aspirated into the air-acetylene flame, and the
absorbances are measured at 460.7 nm.

In a single laboratory, the coefficient of variation at a
strontium concentration of 200 µg was 1.0%.

5.34. Procedures for Thallium

Thallium is conveniently determined in the air-acetylene flame at a wavelength of 276.8 nm. Both the HCL and the FDL are available as sources of thallium resonance radiation. The latter gives higher light output and longer life than the former. Alternate wavelengths are 377.6 nm and 238.0 nm. Use of the nitrous oxide-acetylene flame results in a fourfold loss in sensitivity. The volatility of thallium compounds limits the ashing temperature to 500°C. Hence, the furance technique suffers from matrix interferences, and it frequently requires the method of standard additions to compensate for these interferences.

5.34.1. U.S. EPA Methods 279.1 and 279.2 for Thallium

Method 279.1 is applicable to the determination of thallium in waste water at concentrations ranging from 1 mg/L to 20 mg/ L. Samples are prepared by the procedures described in Section 5.2.1, but the addition of hydrochloric acid is omitted. Samples and standards are aspirated into the air-acetylene flame, and the absorbances are measured at 276.8 nm. The calibration curve is prepared from the absorbances of the standards, and the thallium content of the samples is determined by direct comparison.

In a single laboratory, using mixed industrial-domestic effluent spiked at thallium levels of 0.6 mg/L, 3 mg/L, and 15 mg/L, the coefficients of variation were 3%, 2%, and 0.3%, respectively. The corresponding recoveries were 100%, 98%, and 98%.

Method 279.2 describes the furnace technique for the determination of thallium in waste water at concentrations ranging from 5 µg/L to 100 µg/L. The samples are prepared by the procedures in Section 5.2.1. The absorbances of replicate 20-µL injections of samples and spiked samples are measured at 276.8 nm. The following time-temperature program is recommended: dry, 30 s at 125°C; ash, 30 s at 400°C; atomize, 10 s at 2400°C. Background corrections, continuous argon purge, and nonpyrolytic graphite are recommended.

5.35. Procedures for Tin

The determination of tin by flame atomic absorption spectrometry is made with significantly poorer sensitivity than
that with most other metals can be determined. The measurements are usually made in the nitrous oxide-acetylene flame
at a wavelength of 286.3 nm. Alternate wavelengths are
224.6 nm, 235.5 nm, 270.6 nm, 303.4 nm, 254.7 nm, 219.9 nm,
300.9 nm, and 233.5 nm. The air-acetylene flame can be used
with equal sensitivity to but more interferences than the nitrous oxide-acetylene flame. The sensitivity is improved in
the cooler flames, air-hydrogen and argon-hydrogen, but more
interferences are encountered. Tin can be determined as the
gaseous hydride with significant improvements in the sensitivity. The furnace technique is capable of even greater improvements in sensitivity, but the volatility of tin compounds
limits the ashing temperature which results in more potential
for matrix interferences. The volatility of some of the
interferences can be increased by treatment with ammonia.

5.35.1. U.S. EPA Methods 282.1 and 282.2 for Tin

Method 282.1 describes the determination of tin in waste
water in the range of from 10 mg/L to 300 mg/L. The samples
are prepared by the procedures in Section 5.2.1, and the prepared samples and standards are aspirated into the nitrous
oxide-acetylene flame. The absorbances are measured at
386.3 nm. The calibration curve is prepared from the absorbances of the standards, and the tin content of the samples is
determined by direct comparison.

In a single laboratory, using mixed industrial-domestic
effluent spiked at tin concentrations of 4 mg/L, 20 mg/L, and
60 mg/L, the coefficients of variation were 6%, 3%, and 1%,
respectively. The corresponding recoveries were 96%, 101%,
and 101%.

Method 282.2 describes the furnace technique for the determination of tin in waste water at concentrations ranging from
20 µg/L to 300 µg/L. The samples are prepared by the procedures in Section 5.2.1. The determination of tin is made by

the method of standard additions. If it can be shown to be
applicable, direct comparison may be used. Replicate 20-µL
aliquots of the prepared samples and prepared samples spiked
with known amounts of tin are injected into the graphite furnace, and the absorbances are measured at 224.6 nm using the
following time-temperature program: dry, 30 s at 125°C; ash,
30 s at 600°C; atomize, 10 s at 2700°C. Background corrections, continuous flow argon, and nonpyrolytic graphite are
recommended.

5.36. Procedures for Titanium

The sensitivity with which titanium can be determined is
poorer than that for most other metals. Titanium is determined in the rich nitrous oxide-acetylene flame at 365.3 nm.
There are several alternate wavelengths of almost equal sensitivity: 364.3 nm, 320.0 nm, 363.6 nm, 335.5 nm, 375.3 nm,
334.2 nm, 399.9 nm, and 390.0 nm. Numerous elements interfere
with the determination of titanium by enhancing the sensitivity.
The addition of potassium chloride has been recommended to
standardize these effects. Enhancement by fluoride is corrected for by making all standard and sample solutions 0.1 M
in ammonium fluoride.

5.36.1. U.S. EPA Methods 283.1 and 283.2 for Titanium

Method 283.1 describes the sample preparation and flame
atomic absorption techniques for the determination of titanium
in waste waters at concentrations ranging from 5 mg/L to
100 mg/L. The samples are prepared as follows.

To 100 mL of the well-mixed sample, add 2 mL of concentrated sulfuric acid and 3 mL of concentrated nitric acid.
Cover with a watch glass, and heat on a hotplate adjusted to a
temperature that will cause a gentle reflux action to occur.
Add additional nitric acid as needed to complete the digestion.
When solubilization is complete, remove the watch glass, and
continue heating until dense, white fumes of SO_3 are evolved.
Cool, add 1 mL of 1:1 nitric acid, and dilute to 100 mL with
high-purity water. Treat each prepared sample and each 100 mL
of titanium standard with 2 mL of potassium chloride solution
(95 g/L). Aspirate the treated samples and standards into a

rich nitrous oxide-acetylene flame, and the absorbances are measured at 365.3 nm. Prepare the calibration curve from the absorbances of the standards and determine the titanium content of the samples by direct comparison.

In a single laboratory, using a mixed industrial-domestic effluent spiked at titanium levels of 2 mg/L, 10 mg/L, and 50 mg/L, the coefficient of variation were 4%, 1%, and 0.8%, respectively. Recoveries for these spiked samples were 97%, 91%, and 88%, respectively.

Method 283.2 is applicable to the determination of titanium in waste water samples at levels ranging from 50 μg/L to 500 μg/L. The samples are prepared by the procedures described above in method 283.1. For every sample matrix analyzed, it is necessary to demonstrate that the method of standard additions is not needed in order to use direct comparison for the determination of titanium. The absorbance of 20-μg/L injections of the samples and spiked samples are measured, in replicate at 365.4 nm. The following time-temperature program is recommended: dry, 30 s at 125°C; ash, 30 s at 1400°C; atomize, 15 s at 2800°C. Background corrections, continuous argon purge, and pyrolytic graphite are also recommended.

5.37. Procedures for Vanadium

Vanadium is determined in a rich nitrous oxide-acetylene flame using the triplet at 318.3-318.4-318.5 nm. The sensitivity is less than that of most other metals. Aluminum enhances the sensitivity, and corrections for this effect are made by adding aluminum to both samples and standards. Alternate wavelengths are 306.6 nm, 306.0 nm, 305.6 nm, 320.2 nm, and 390.2 nm. Even though the thermal stability of vanadium allows ashing temperatures as high as 1600°C, there are interferences when the furnace technique is employed. The method of standard additions is frequently used when such interferences are encountered.

5.37.1. U.S. EPA Methods 286.1 and 286.2 for Vanadium

Method 286.1 describes the determination of vanadium in waste water samples at concentrations ranging from 2 mg/L to 100 mg/L. The samples are prepared by the procedures in

Section 5.2.1. The prepared samples and the standards are
treated with 2 mL (per 100 mL of sample or standard) of the
same aluminum nitrate solution used in the determination of
molybdenum, Section 5.21.1, method 246.1. The treated samples and standards are aspirated into a rich nitrous oxide-
acetylene flame. The absorbances at the 318.3-318.4-318.5 nm
triplet are recorded. Using the absorbances of the standards
for the calibration curve, the vanadium levels in the samples
are determined by direct comparison.

In a single laboratory, using a mixed industrial-domestic
effluent spiked at vanadium levels of 2 mg/L, 10 mg/L, and
50 mg/L, the coefficients of variation were 5%, 1%, and 0.4%,
respectively. The corresponding recoveries were 100%, 95%,
and 97%.

Method 286.2 describes the furnace technique for the determination of vanadium in waste waters. The optimum concentration range for this method is from 10 µg/L to 200 µg/L. The
procedures in Section 5.2.1 are used to prepare the samples.
The vanadium content of the samples is determined by standard
additions unless it can be shown that there are no interferences. If this is demonstrated, direct comparison may be used.
The absorbances of replicate 20-µL injections of the spiked
and unspiked samples are measured at 318.4 nm using the following time-temperature program: dry, 30 s at 125°C; ash,
30 s at 1400°C; atomize, 15 s at 2800.°C. Background corrections, continuous purge with argon, and pyrolytic graphite are
recommended.

5.37.2. Canadian DOE Methods 23001 and 23002 for Vanadium

Method 23001 is applicable to the determination of vanadium
in a wide variety of waters, including surface waters and
waste waters. The samples are prepared by the procedures described in Section 5.2.2, and the prepared samples and standards are aspirated into the nitrous oxide-acetylene flame.
The absorbances are measured at 318.4 nm, and the vanadium
content of the samples is determined by direct comparison
using the absorbances of the standards for the calibration
curve.

Method 23002 describes a chelation-extraction procedure
for the determination of vanadium at concentrations ranging
from 0.5 µg/L to 20 µg/L. Samples are prepared by the

procedures in Section 5.2.2. The prepared samples and standards are then treated as follows.

Pipette a 150-mL aliquot into a 200-mL volumetric flask and
adjust to the pH range of 2.5 to 3.5. To each flask, add
0.5 mL and 1% bromine water and warm the contents on a water
bath until the color of the bromine disappears. Cool, add
5.0 mL of 1% aqueous cupferron, and mix well. Add 3.0 mL of
n-butyl acetate to each and shake the flasks for 5 min. Allow
the phases to separate and float the organic phases up into
the necks of the flasks by slowly adding water down the sides.
Aspirate the organic phases into the nitrous oxide-acetylene
flame, and measure the absorbances at 318.4 nm. Determine
the vanadium content of the samples by direct comparison using
the absorbances of the standards for the calibration curve.

5.38. Procedures for Zinc

Zinc is conveniently determined in the air-acetylene flame
at a wavelength of 213.9 nm. Use of the nitrous oxide-
acetylene flame results in a threefold loss in sensitivity.
The volatility of zinc compounds limits the ashing step to a
temperature of 400°C. Hence, matrix interferences are frequently encountered with the furnace technique, and the method
of standard additions is frequently used to minimize the effects of these interferences.

5.38.1. U.S. EPA Methods 289.1 and 289.2 for Zinc

Method 289.1 is applicable to the determination of zinc in
wastewater samples at concentrations ranging from 0.05 mg/L to
1 mg/L. The samples are prepared by the procedures described
in Section 5.2.1. For samples with zinc levels below 0.05 mg/
L, the chelation-extraction procedure from Section 3.2.1 may
be used for preconcentration. When chelation-extraction is
employed, both the samples and standards must be carried
through the procedure. The samples and standards are aspirated
into the air-acetylene flame, and the absorbances at 213.9 nm
are measured. Using the absorbances of the standards for the
calibration curve, the zinc content of the samples is determined by direct comparison.

Some seven dozen laboratories analyzed natural water spiked

with six synthetic concentrates containing varying amounts of aluminum, cadmium, chromium, copper, iron, manganese, lead, and zinc. For samples spiked at the 300-ppb zinc level, the mean results showed a coefficient of variation of 33% and a bias of 1%. Samples spiked at the 60-ppb level showed a coefcient of variation of 44%. The mean results were within 10% of the theoretical values at the 60-ppb zinc level.

Method 289.2 describes the furnace technique for the determination of zinc in waste water at levels between 0.2 µg/L and 4 µg/L. The samples are prepared in accord with the procedures in Section 5.2.1, and the absorbances of replicate 20-µL injections of the prepared samples and samples spiked with the zinc standards are measured at 213.9 nm using the following time-temperature program: dry, 30 s at 125°C; ash, 30 s at 400°C; atomize, 10 s at 2500°C. If it can be demonstrated that the method of standard additions is not needed, the zinc content of the samples may be determined by direct comparison. Background corrections, continuous argon purge, and nonpyrolytic graphite are recommended.

5.38.2. Canadian DOE Methods 30004 and 30005 for Zinc

Method 30004 is applicable to the determination of zinc in a wide variety of water samples. The samples are prepared by the procedures in Section 5.2.2, and the prepared samples and standards are aspirated into the air-acetylene flame. The absorbances at 213.8 nm are recorded, and the zinc content of the samples is determined by direct comparison using the absorbances of the standards for the calibration curve.

Method 30005 describes the chelation-extraction procedure for the determination of low levels of zinc in a wide variety of waters. The samples are prepared by, and the prepared samples and standards are extracted by, the procedures in Section 5.2.2. The organic phases are aspirated into the air-acetylene flame. The absorbances are measured at 213.8 nm, and the zinc content of the samples is determined by direct comparison using the absorbances of the standards for the calibration curve.

In a single laboratory, the coefficient of variation at a zinc level of 10 ppb was 1.4% using the chelation-extraction procedure.

The methods described above were abstracted from the U.S.

EPA's <u>Methods</u> <u>for</u> <u>Chemical</u> <u>Analysis</u> <u>of</u> <u>Water</u> <u>and</u> <u>Wastes</u>,[9] and
from the Canadian DOE's <u>Analytical</u> <u>Methods</u> <u>Manual</u>.[13]   In addi-
tion, the EPA Manual,[8] the American Public Health Association's
(APHA) Standard Methods,[17] the American Society for Testing
and Materials (ASTM) Annual Book of Standards,[15, 16] U.S. Geological Survey (USGS) Manuals,[162, 163] the Association of
Official Analytical Chemists (AOAC)Official Methods,[164] and
even American National Standard Institute (ANSI) Standards[165]
have been identified as sources of approved methods for regulatory compliance monitoring by atomic absorption spectrometry.
The guidelines establishing test procedures for the analysis
of pollutants under the National Pollution Discharge Elimination System (NPDES) program have approved atomic absorption
procedures from all but the first two items cited above, and
the inorganic chemical sampling and analytical requirements
for the national primary and secondary drinking water regulations have relied upon references 8 and 17 above as sources of
approved atomic absorption procedures.

6. METHODS FOR COMPLIANCE MONITORING OF LIQUID WASTES, SOLID WASTE, SLUDGES, SEDIMENTS, AND SOILS

This chapter contains methods that are to be used in evaluating wastes for establishing their inclusion or exclusion as hazardous wastes under PL 94-580, the Resource Conservation and Recovery Act (RCRA). These methods are contained in Test Methods for Evaluating Solid Wastes,[14] and they are applicable to the determination of metals for identifying both the composition of the waste and its "extraction procedure toxicity" (EP toxicity). Also included in this section are the U.S. Environmental Protection Agency (EPA) methods for the determination of heavy metals in sewage sludge[61] and for the determination of mercury in sludges, sediments, and soils,[9] and the Canadian Department of the Environment (DOE) methods for the determination of heavy metals in sediments and soils.[13] The latter methods have undergone some degree of inter-laboratory testing. The methods contained in Test Methods for Evaluating Solid Wastes, however, are "state of the art methodologies for conducting such tests"; ie, many of the methods presented in this manual have not been fully evaluated using materials characteristic of the wastes regulated under RCRA.

6.1. Sample Collection and Preservation

The various aspects of sample collection and preservation are presented in Chapter 2 and they are not repeated here. Of specific importance are Sections 2.1.5, 2.2.3, 2.3.4, 2.4.6, and 2.5.3.

6.1.1. U.S. EPA Procedures

The U.S. EPA procedures for sampling wastes are summarized in Tables 2.1, 2.2, and 2.5. The samples may be collected in borosilicate glass, polyethylene, polypropylene, or Teflon® bottles with screw caps. The waxed paper liners from the rigid plastic caps of the glass bottles should be replaced with inert plastic liners. The bottles, caps, and liners are precleaned by the following procedure:

1. Thoroughly scrub with detergent and water.

2. Rinse with a solution of 1 part concentrated nitric acid to 1 part water.

3. Rinse with water.

4. Rinse with a solution of 1 part hydrochloric acid to 1 part water.

5. Rinse with water.

6. Rinse with deionized, distilled water.

7. Dry the plastics at 50°C, the glassware at 105°C.

Unlike water and wastewater samples, the U.S. EPA does not allow the addition of preservatives to samples of solid and/or liquid wastes. The latter samples are cooled with either ice or dry ice to retard changes and delivered, "as soon as practicable--usually within 1 or 2 days after sampling," to the laboratory for analysis.

6.1.2. Canadian DOE Procedures

Sediment samples may be collected using an Ekman dredge, a grab sampler, or a core sampler depending on the requirements of the site. Immediately after collection, samples should be drained and stored frozen at -20°C in polypropylene bags or bottles. Storage times of up to 6 months are permitted for samples treated in this manner.

6.2. Sample Preparation

As described in Chapter 3, the samples must be prepared by appropriate chemical and/or physical procedures for introduction into the atomic absorption spectrometer. The regulatory agencies have recommended, and in some cases required, specific sample preparation procedures for this purpose.

Table 6.1. Maximum Concentration of Contaminants
for EP Toxicity [Code of Federal Register
(CFR) 261.24]

EPA hazardous waste number	Contaminant	Maximum concentration (mg/L)
D004	Arsenic	5.0
D005	Barium	100.0
D006	Cadmium	1.0
D007	Chromium	5.0
D008	Lead	5.0
D009	Mercury	0.2
D010	Selenium	1.0
D011	Silver	5.0

6.2.1. U.S. EPA Extraction Procedure Toxicity, Method 1310

The extraction procedure (EP) is designed to simulate the
leaching a waste will undergo if disposed of in an improperly
designed landfill. It is a laboratory test in which a representative sample of a waste is extracted with distilled water
maintained at pH 5 using acetic acid. The extract obtained
from the EP (the EP extract) is then analyzed to determine
whether any of the thresholds established for arsenic, barium,
cadmium, chromium, lead, mercury, selenium, and/or silver have
been exceeded. These thresholds are listed in Table 6.1. If
the EP extract contains any one of these metals in an amount
equal to or exceeding the levels cited in Table 6.1, the waste
possesses the characteristic of extraction procedure toxicity,
and it is identified as a hazardous waste.

The extraction procedure consists of five steps:

1. Separation procedure. A waste containing unbound liquid
   is filtered. If the solid phase is less than 0.5% of
   the waste, the solid phase is discarded, and the filtrate
   is analyzed for the metals. If the waste contains more
   than 0.5% solids, the solid phase is extracted and the
   liquid phase is retained for later use.

2. Structural integrity procedure/particle size reduction.
   Prior to extraction, the solid material must either pass
   through a 9.5-mm (0.375-in.) standard sieve or have a
   surface area per gram of 3.1 cm^2. If the solid consists
   of a single piece, it must be subjected to the structural
   integrity procedure. This procedure is used to demonstrate the ability of the waste to remain intact after
   disposal. If the waste does not meet one of these conditions, it must be ground to pass through the 9.5-mm
   sieve.

3. Extraction of solid material. The solid material from
   step 2 is extracted for 24 h in an aqueous medium maintained at or below pH 5 with 0.5 N acetic acid. The pH
   is maintained either automatically or manually. Acidification to pH 5 is subject to a specification as to
   the total amount of acid to be added to the system.

4. Final separation of the extraction from the remaining
   solid. After extraction, the liquid to solid ratio is
   adjusted to 20:1, and the mixture of solid and extraction liquid is separated by filtration. The solid is
   discarded, and the liquid phase is combined with the
   filtrate obtained in step 1. This combination is the
   EP extract that is analyzed to determine whether any of
   the thresholds have been exceeded.

5. Analysis of extraction procedure extract. The metals
   are identified and quantitated by the methods specified
   in Sections, 6.4.1, 6.5.1, 6.6.1, 6.8.1, 6.10.1, 6.14.1,
   6.18.2, 6.20.1, 6.21.1, and 6.22.1 for antimony,
   arsenic, barium, cadmium, chromium, lead, mercury,
   nickel, selenium, and silver respectively.

The five steps of the extraction procedure are summarized
as a flow chart in Figure 6.1.

6.2.1.1. Separation Procedure   The first step in the extraction procedure is the separation of the waste sample into its
liquid and solid phases. To accomplish this separation, a
representative sample of the waste, collected, preserved, and
stored in the prescribed manner, and weighing at least 100 g,
is subjected to filtration, using pressure when vacuum is in-
adequate for complete separation.

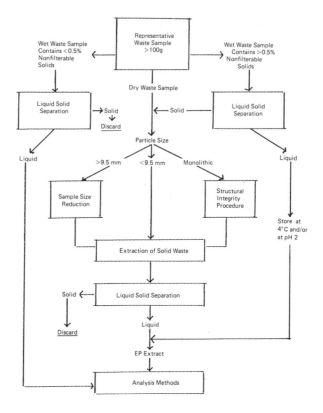

Figure 6.1. Extraction procedure flowchart.

The filter holder must be capable of supporting a 0.45 μm filter membrane and able to withstand the pressure needed to accomplish the separation. Suitable filter holders range from simple vacuum units* to relatively complex systems that can exert up to 75 psi of pressure.† The properties of the waste sample to be filtered determine whether vacuum or pressure filtration is to be employed.

---

*Nalgene model 45-0045, Nucleopore model 410400, Millipore model XX1004700.

†Nucleopore model 420800, Micro Filtration System model 302300, Millipore model YT 30142 HW.

The filter membrane must (1) separate the "free" liquid portion of the waste from any solid matter having a particle size greater than 0.45 μm, (2) not absorb or leach the chemical species for which the waste's EP extract will be analyzed, and (3) not be physically changed by the waste material to be filtered. The following procedure is used to determine whether a filter membrane will absorb or leach analyte:

1. Prepare a standard solution of the analytes.

2. Analyze this standard for the metals of interest.

3. Filter the standard and again analyze for the metals of interest. If the levels of analytes differ from those of the unfiltered standard, the filter membrane leaches or absorbs one or more of the metals.

To determine whether a filter membrane (or a prefilter) is adversely affected by a particular waste, submerge the filter in the waste's liquid phase and examine the filter for swelling, shrinkage, solubility, curling, etc, 48 h later. If any of these physical changes is observed, the filter is unsuitable for use.

Following the manufacturer's directions, assemble the filtration apparatus using preweighed (with a precision of ±10 mg) filter units. For difficult or slow to filter wastes, a filter unit consisting of a 0.45-μm membrane filter, a 0.65-μm membrane filter, a fine glass fiber prefilter, and a coarse glass fiber prefilter can be used.

Allow the waste sample to stand so that the solid phase settles. Wastes slow to settle may be contrifuged. Wet the filter with a small portion of liquid phase from the waste and then transfer the remainder of the sample material to the filtration apparatus. Apply vacuum or 10 to 15 psi pressure until all of the liquid passes through the filter. Stop the filtration when air (or pressurizing gas) passes through the filter unit. If this point is not reached under vacuum or a pressure of from 10 to 15 psi, slowly increase the pressure in 10 psi increments to 75 psi. Stop the filtration when the flow of liquid is no longer observed. The filtrate ("free" liquid portion of the waste sample) should be stored at 4°C for subsequent use.

If the solid phase comprises less than 0.5%* of the waste, discard the material on the filter unit. If, on the other hand, the waste sample is more than 0.5%* solid phase, the material on the filter unit must be subjected to the extraction procedure described in Section 6.2.1.3.

If the extraction procedure is to be carried out, remove the filter unit containing the solid phase of the waste and reweigh it (again with a precision of ±10 mg) immediately. Use the wet weight of the solid phase obtained in this separation procedure to calculate the 16:1 water to solid ratio required for the extraction procedure as follows:

$$W = W_f - W_t$$

where W is the weight of solid phase of the waste used to charge the extractor, W_f is the wet weight of the filter unit containing the solid phase, and W_t is the tare weight of the filter unit. The extractor is also charged with a quantity of deionized water equal to 16 W.

6.2.1.2. Structural Integrity Procedure   The structural integrity procedure (SIP) is used as an approximation of the physical degradation a monolithic waste undergoes in a landfill or when compacted by earth moving equipment. A sample of the waste, cut into a 3.3-cm diameter by 7.1-cm long cylinder, is placed in the sample holder of the structural integrity tester[†] shown in Figure 6.2. For wastes that have

---

*The percentage solids of a waste sample is determined separately. A weighed sample of the waste is filtered through a tared filter unit, and the filter unit with the solids is reweighed after drying to constant weight at 80°C. The percentage solids of the waste is calculated as:

$$\% \text{ solids} = 100 \times \frac{(\text{dry weight of filter unit with solids}) - (\text{tare weight})}{\text{initial weight of sample}}$$

Do not extract solid material that has been dried.

[†]Associated Design and Manufacturing Co., Alexandria, VA 22314.

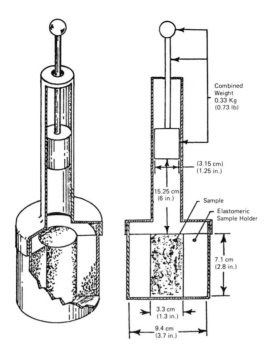

*The elastomeric sample holder is fabricated of material firm
  enough to support the sample.

Figure 6.2. Compaction tester.

been treated by a fixation process, the waste may be cast into
a cylinder of these dimensions and allowed to cure for 30 days
prior to testing. The hammer is raised to its maximum height
and dropped on the cylinder of waste a total of 15 times. The
waste is then removed from the sample holder, any material ad-
hering to the holder is scraped off and added to that removed
from the holder, and the waste material is weighed with a precision of ±10 mg. The waste is then transferred to the
extractor.

6.2.1.3. Solid Material Extraction Procedure   The solid material from the separation procedure (Section 6.2.1.1) must be
evaluated for particle size prior to extraction. If this material has a surface of 3.1 cm^2/g or more, or if this material

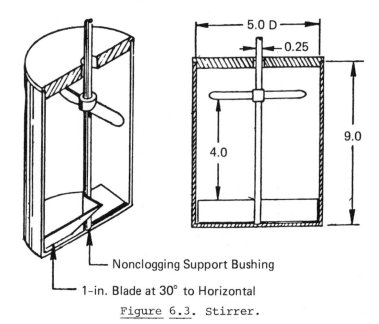

Nonclogging Support Bushing

1-in. Blade at 30° to Horizontal

Figure 6.3. Stirrer.

passes through a 9.5-mm seive, it may be subjected to the extraction procedure without further treatment. If, on the other
hand, neither of these criteria is met, the solid material is
cut, crushed, or ground until it passes through a 9.5-mm sieve.

The solid material from either the separation procedure or
the structural integrity procedure is weighed and immediately
placed in an extractor with 16 times its weight of deionized
water. There are two types of acceptable extractors: stirrers and tumblers. Each must impart sufficient agitation to
the mixture not only to prevent stratification of the sample
and extraction fluid but also to insure that all sample surfaces are continuously brought into contact with well-mixed
extraction fluid. Stirrers consist of a container in which
the waste-extraction fluid mixture is agitated by spinning
blades as shown in Figure 6.3. Tumblers agitate the waste-
extraction fluid mixture by turning the bottle containing the
mixture end over end (Figure 6.4) or by turning the bottle on
its side and rotating it on its cylindrical axis (Figure 6.5).
Tumblers are preferred to stirrers for the extraction of monolithic wastes.

The pH of the waste-extraction fluid mixture should be

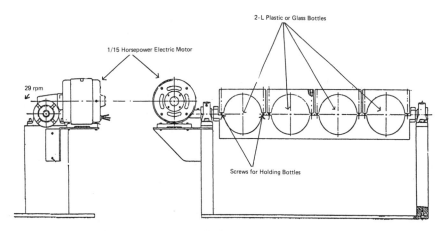

Figure 6.4. Tumbler.

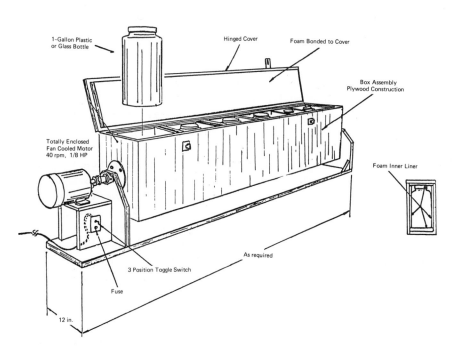

Figure 6.5. Tumbler.

measured and, if found to be above 5.0, 0.5 N acetic acid
should be added to the contents of the extractor until a pH of
5.0 ± 0.2 is obtained. The pH of the mixture should be checked
and adjusted, as needed, at 15-, 30-, or 60-min intervals* during the first 6 h of the extraction. If the pH does not have
to be adjusted more than 0.5 units, move on to the next longer
time interval between checks and adjustments. The aggregate
amount of 0.5 N acetic acid added to the waste-extraction fluid
mixture should not exceed 4 mL of acid per gram of solids. The
extraction should be continued for 24 h at a temperature between 20° and 40°C. If at the end of the 24-h extraction
period the pH of the solution is not below 5.2 and the maximum
amount of acid (4 mL per gram of solids) has not been added,
the pH should be adjusted to 5.0 ± 0.2, and the extraction
should be continued for an additional 4 h. The pH should be
checked and adjusted, if necessary, at 1-h intervals during
this time.

   At the end of the extraction period, deionized water is
added to the contents of the extractor in an amount calculated
to give a 20:1 extraction fluid to solids ratio. The amount
of water to be added, V, is calculated as:

$$V = 20W - 16W - A$$

where W is the weight of solid phase of the waste used to
charge the extractor and A is the total volume of 0.5 N acetic
acid added during the extraction to maintain the pH at 5. W
and A are expressed in grams and milliliters, respectively.

6.2.1.4. Procedure for the Final Separation of Extract from
Remaining Solids   The final separation of the extraction
fluid from the solids is accomplished by the same filtration
processes as those described in Section 6.2.1.1. The liquid
phase from the extraction procedure and the "free" liquid portion from the initial separation of the waste sample into its
liquid and solid phases are combined. This combination of the
two liquids is the EP extract that is analyzed for metals by

---

*An automatic pH controller such as the Type 45-A manufactured
  by Chemtrix, Inc., Hillsboro, Oregon 27123, is recommended
  for maintaining the pH in stirrer-type extractors.

atomic absorption spectrometry as described in subsequent
sections of this chapter.

6.2.2. U.S. EPA Procedure for Oil-, Grease-, or Wax-Containing Wastes, Method 3030[14]

Samples of waste containing substantial amounts of organic
materials require vigorous digestion prior to the determination of metals by atomic absorption spectrometry. The sulfuric acid-nitric acid digestion presented in Section 3.1.1.3
is applicable to the analysis of oil-, grease-, or wax-containing wastes for the elements arsenic, mercury, selenium,
cadmium, chromium, silver, lead, and barium.

6.2.3. U.S. EPA Procedure for Industrial Effluents[60]

The guidelines for the preparation of industrial effluent
samples prior to the determination of metals by conventional
flame atomic absorption spectrometry recommend the nitric acid
digestion described in Section 3.1.1.2. With the exceptions
of antimony and beryllium, samples of industrial effluent to
be analyzed by furnace techniques should be prepared as
follows:[60]

> Transfer 100 mL of well mixed sample to a 250 mL beaker,
> add 3 mL of concentrated, redistilled nitric acid
> and 5 mL of 30% hydrogen peroxide, and heat at 95°
> for one hour, or until the volume is reduced to less
> than 50 mL. Cool, and dilute to 50 mL with de-
> ionized water.

The sample preparation procedures for antimony and beryllium
analysis by furnace techniques are the same as those used
when conventional flame techniques are employed.

6.2.4. U.S. EPA Procedures for Municipal Sludges[61]

The nitric acid-hydrogen peroxide digestion procedure described in Section 3.1.1.3 is applicable to the preparation
of sludge samples prior to analysis by atomic absorption
spectrometry. With the exception of mercury, this digestion
is suitable for both flame and furnace analyses of all other

priority pollutants: antimony, arsenic, beryllium, cadmium, chromium, copper, lead, nickel, selenium, silver, thallium, and zinc.

The efficiency of this digestion procedure was evaluated using the U.S. EPA's Municipal Digested Sludge as the reference material in an independent round robin.[166] Sixteen participating laboratories analyzed five replicate samples of this material for up to 19 different metals. Some 85% of the individual results were within two standard deviations of the reference values, and the coefficients of variation, based on the grade means for each element, were less than 50% for 14 of the 19 metals, and less than 20% for eight of the 19 metals.

6.2.5. Canadian DOE Bomb Digestion Procedure for Sediments

This method is applicable to the determination of aluminum, barium, beryllium, cadmium, calcium, chromium, cobalt, copper, iron, lead, lithium, magnesium, manganese, molybdenum, nickel, potassium, silicon, sodium, strontium, vanadium, and zinc in sediments. Some general applications of the bomb digestions were described in Section 3.1.1.8. Sediment samples are prepared as follows.

Samples collected and stored pursuant to Section 6.1.2 are thawed and air dried at room temperature. Representative subsamples obtained by coning and quartering are ground to pass through a 270-mesh sieve. Replicate 100-mg portions of this fine powder are weighed with a precision of ±0.2 mg, transferred to the Teflon® cups of the pressure decomposition vessels, and treated with 4.0 mL of nitric acid, 1.0 mL of perchloric acid, and 6.0 mL of hydrofluoric acid. The vessels are assembled, sealed, and placed in an oven maintained at 140°C for 3.5 h. After this time, the bombs are removed from the oven, cooled, and opened. The contents of the Teflon® cups are quantitatively transferred to 125-mL bottles containing 4.8 g of boric acid and 30 mL of deionized water. The contents of the bottles are shaken until all of the material dissolves, then transferred to 100-mL volumetric flasks for volume adjustments, and finally returned to polypropylene bottles within 2 h for storage. Samples prepared in this manner may be stored for up to 40 days prior to the determination of the metals cited above by atomic absorption spectrometry.

6.2.6. Canadian DOE Open Digestion Procedure for Sediments

This method is applicable to the determination of aluminum, cadmium, cobalt, copper, iron, lead, manganese, molybdenum, nickel, vanadium, and zinc in soils and sediments. Freeze-dried samples are preferred to frozen and subsequently air-dried samples. Initial coarse (20 mesh) sieving is used to remove extraneous debris, and subsequent grinding of a sub-sample, obtained by coning and quartering, is used to obtain a representative material for analysis. Portions of this material, weighing between 1 and 3 g, are transferred to Teflon® beakers and treated with 15 mL of concentrated nitric acid. The beakers are heated on a hotplate covered with a sheet of asbestos. After the contents have boiled for 2 min, 10 mL of perchloric acid are added and the heating is continued until a white paste remains in the beakers. (CAUTION: perchloric acid should be boiled only in an approved hood. The beakers should not be permitted to boil dry.) Ten milliliters of hydrofluoric acid are added to the white pastes in the beakers, and the heating is continued until the residues dissolve. The hydrofluoric acid is boiled off, and the contents of the beakers are dissolved in 5 mL of concentrated hydrochloric acid and 20 mL of deionized water with gentle heating. These solutions are quantitatively transferred to 100-mL volumetric flasks and brought to volume with deionized water.

The precision and accuracy of this digestion procedure has been evaluated. In a single laboratory, nine replicates of a sediment from Lake Ontario were prepared by this method and analyzed for the 11 metals cited above. The coefficients of variation were below 2.5% for five of the metals and below 5% for seven of the metals; none of the coefficients of variation exceeded 7.5%. At the beginning of digestion, sediments from Lake Ontario, Lake Huron, and Lake Erie were spiked to 100 µg with all of the elements cited above except aluminum and iron. Recoveries ranged from 96% to 103%. The U.S. Geological Survey (USGS) standards W-1 and BCR-1 were prepared for atomic absorption spectrometry by this procedure. The values determined for aluminum, iron, manganese, cobalt, copper, lead, zinc, and vanadium agreed with the certified values to within ±5%.

6.2.7. Canadian DOE Acid Extraction Procedure for Sediments

This method is applicable to the determination of aluminum,

cadmium, chromium, cobalt, copper, iron, lead, manganese, nickel, and zinc in sediments. Metals determined in samples prepared by this extraction procedure are identified as "non-residual," which is defined as the fraction of the metal that is not part of the silicate matrix of the rock from which the sediment is derived. This includes metals absorbed on the sediment particles, complexed by and absorbed on organic matter, and in the form of insoluble salts.

Freeze-dried samples are preferred to frozen and subsequently air-dried samples. Aggregates formed during air drying are crushed, and the coarse material is removed by sieving (20 mesh). The material passing through the 20-mesh sieve is subjected to further sieving on an 80-mesh sieve. The material passing through the latter is coned and quartered, and a representative 10-g portion is weighed with a ±10-mg precision and placed in a 125-mL wide mouth, polypropylene bottle. The contents of the bottle are treated with 100 mL of 0.5 N hydrochloric acid. The bottle is tightly capped and shaken at room temperature overnight (16 h). After this time, the contents of the bottle are filtered through a 0.45-μm cellulose acetate filter. The filter and the solid phase are discarded, and the filtrate is retained for the determination of the metals cited above by atomic absorption spectrometry.

In a single laboratory, the determination of metal levels in replicate portions of a sediment extracted as described above showed the coefficients of variation indicated: Al, 6640 mg/kg ± 4.0%; Cd, 1.4 mg/kg ± 8.8%; Co, 10 mg/kg ± 0.7%; Cr, 14 mg/kg ± 2.9%; Cu, 40 mg/kg ± 1.0%; Fe, 14,300 mg/kg ± 2.0%; Mn, 1240 mg/kg ± 3.3%; Ni, 30 mg/kg ± 2.2%; Pb, 44 mg/kg ± 4.0%; Zn, 70 mg/kg ± 0.5%.

6.3. Procedures for Aluminum

The general conditions for the determination of aluminum are described in Section 5.3.

6.3.1. Canadian DOE Method 13050 for Aluminum

Method 13050 is applicable to the determination of aluminum in sediment samples prepared by the acid extraction procedure (Section 6.2.7), the open digestion procedure (Section 6.2.6),

or the bomb digestion procedure (Section 6.2.5). The hydrofluoric and boric acids used in the bomb digestion procedure
depress the sensitivity of the aluminum determination. Hence,
it is necessary to incorporate similar concentrations of these
acids into the standards. The samples and standards are
aspirated into a reducing nitrous oxide-acetylene flame, and
the absorbances are measured at a wavelength of 309.3 nm. The
aluminum levels of the samples are determined by direct comparison. At aluminum levels of 5000 mg/kg, 10,000 mg/kg, and
15,000 mg/kg the coefficients of variation were 7.1%, 6.5%,
and 10.1%, respectively. Recoveries of spikes added to a
series of samples averaged 99%.

6.4. Procedures for Antimony

Section 5.4 contains the description of the general conditions for the determination of antimony.

6.4.1. U.S. EPA Methods 7040 and 7041 for Antimony

Methods 7040 and 7041 have been approved for the determination of antimony in waste samples prepared by the digestion
procedure described below or in EP extracts obtained by following the procedures described in Section 6.2.1.

For the direct aspiration technique, method 7040, 100 mL of
well-mixed sample or 100 mL of EP extract is placed in a
250-mL beaker and 3 mL of concentrated nitric acid is added.
The contents of the beaker are heated to near dryness on a
hotplate. After cooling, another 3 mL of nitric acid are
added, the beaker is covered, and the contents are heated at a
gentle reflux, with possible further additions of 3-mL increments of acid, until a light colored solution is obtained.
The contents of the beaker are evaporated to near dryness,
cooled, treated with 1 mL of 1:1 hydrochloric acid, and
warmed briefly to dissolve any residue. The contents of the
beaker are quantitatively transferred to a 100-mL volumetric
flask and diluted to the mark with 0.2% v/v nitric acid.

The antimony content of the digested waste samples or EP
extracts is determined by the method of standard additions
using either the flame or the furnace technique.

For the flame technique, method 7041, 2-mL aliquots of the

digested sample or extract are pipetted into each of four
10-mL volumetric flasks. No addition is made to the first
flask, 5 mL of 50-µg/L antimony standard are added to the second flask, 5 mL of 75-µg/L standard to the third flask, and
5 mL of 100-µg/L standard to the fourth. The spiked and unspiked samples are aspirated into a lean air-acetylene flame,
and the absorbances are measured at 217.6 nm. (If lead is
present, these measurements are made at 231.1 nm.) The working range of the flame technique is from 1 to 40 mg/L.

The furnace technique is applicable to the determination of
antimony at levels 100 to 1000 times lower than the working
range of the flame technique. Replicate 20-µL injections of
the spiked and unspiked samples are made into a nonpyrolytic
graphite furnace tube. Absorbance measurements are made at
217.6 nm using continuous flow purge gas and the following
furnace parameters: dry, 30 s at 215°C; ash, 30 s at 800°C;
atomize, 10 s at 2700°C. If chloride matrix problems are encountered, the addition of 5 ppm ammonium nitrate solution is
sometimes helpful.

6.5. Procedures for Arsenic

The general conditions for the determination of arsenic
are presented in Section 5.5.

6.5.1. U.S. EPA Methods 7060 and 7061 for Arsenic

Methods 7060 and 7061 have been approved for the determination of the arsenic content of waste samples and for the determination of the arsenic levels in EP extracts. Either the
furnace technique or the gaseous hydride technique may be
used. Each, however, requires a somewhat different procedure
for the preparation of the waste sample.

For the furnace technique, method 7060, a 100-g sample of
the well-mixed waste or 100 mL of EP extract is transferred to
a 250-mL beaker, treated with 2 mL of 30% hydrogen peroxide
and 1 mL of concentrated nitric acid, and heated on a 95°C
hotplate until the volume is reduced to less than 50 mL. It
may be necessary to add more peroxide and acid for complete
digestion. The digested sample or extract is cooled, transferred to a 50-mL volumetric flask, and diluted to the mark
with high-purity water.

A 25-mL aliquot of the digested and diluted sample or extract is transferred to a second 50-mL volumetric flask, treated with 5 mL of 1% nickel nitrate solution, and diluted to the mark. The arsenic content of the final solution is determined by the method of standard additions.

Standard additions of 5 mL of 100 µg As per liter, 5 mL of 75 µg As per liter, 5 mL of 50 µg As per liter, and no addition are made to 2-mL aliquots of the final solution contained in 10-mL volumetric flasks. The contents of each flask are treated with 0.3 mL of 1% nickel nitrate solution and 0.3 mL of 30% hydrogen peroxide and brought to volume. Replicate 20-µL injections are made into the graphite furnace and absorbance measurements are made with continuous purge gas flow at a wavelength of 193.7 nm using the following program: dry, 30 s at 125°C; ash, 30 s at 1100°C; atomize, 10 s at 2700°c. The working range of the furnace technique is from 5 µg to 100 µg As per liter.

The working range of the gaseous hydride technique, method 7061, is from 2 µg to 20 µg As per liter. A 50-g sample of the waste or 50 mL of the EP extract is transferred to a 100-mL beaker and treated with 10 mL of concentrated nitric acid and 12 mL of dilute (18 N) sulfuric acid. The contents of the beaker are heated until dense, white fumes of SO_3 are evolved. Charring of the sample is prevented by maintaining an excess of nitric acid during the heating stage until a colorless or light straw solution is obtained prior to the evolution of the SO_3. The colorless solution is cooled, treated with 25 mL of high-purity water, and again heated until dense, white fumes of SO_3 are evolved. The digested sample is cooled, transferred to a 100-mL volumetric flask, treated with 40 mL of concentrated hydrochloric acid, and diluted to the mark with high-purity water. The arsenic content of the digested sample is determined by the method of standard additions.

To four 10-mL aliquots of digested sample contained in separate 50-mL volumetric flasks additions corresponding to 0 µg, 375 µg, 500 µg, and 625 µg are made. The contents of the flasks are diluted to 50 mL with 0.15% v/v nitric acid. A 25-mL aliquot of the contents of the flask is transferred to the reaction vessel of the hydride generator, treated with 1 mL of 20% w/v potassium iodide solution and 0.5 mL of a solution made by dissolving 100 g of stannous chloride in 100 mL of concentrated hydrochloric acid. After an incubation period of 10 min, the reaction vessel is connected to the gas handling system of the hydride generator, and 1.5 mL of zinc slurry are injected into the reaction vessel. Absorbance measurements are made at a wavelength of 193.7 nm in the argon-hydrogen flame.

6.6. Procedures for Barium

The general conditions for the determination of barium are presented in Section 5.6.

6.6.1. U.S. EPA Methods 7080 and 7081 for Barium

Methods 7080 and 7081 contain approved procedures for the preparation of waste samples and EP extracts prior to the determination of barium by flame or furnace atomic absorption spectrometry. The working range for the former is from 1 mg/L to 20 mg/L, and the latter is useful for the determination of barium concentrations in the range of from 10 µg/L to 200 µg/L.

The procedures used to prepare waste samples and EP extracts for flame or furnace atomization are essentially the same except that the latter avoids the addition of chloride ion in all forms. A 100-mL sample of the well-mixed waste or a 100-mL sample of the EP extract is transferred to a 250-mL beaker, treated with 3 mL of concentrated nitric acid, and slowly evaporated to near dryness by heating it below the boiling point on a hotplate. The contents of the beaker are cooled, treated with an additional 3 mL of nitric acid, covered with a watch glass, and heated gently. Additional 3-mL increments of acid may be made and the heating continued until a colorless or pale yellow solution is obtained. This solution is evaporated to near dryness and allowed to cool.

For barium determinations made by the flame technique, method 7080, the contents of the beaker are treated with 2 mL of 1:1 hydrochloric acid and warmed to dissolve any residue. The watch glass and inner walls of the beaker are washed down with high-purity water, and the contents of the beaker are quantitatively transferred to a 100-mL volumetric flask. If the air-acetylene flame will be used to atomize the barium, 4.7 mL of lanthanum chloride (25 g of lanthanum oxide dissolved in 250 mL of concentrated hydrochloric acid and diluted to 500 mL with high-purity water) solution is added to the

flask. The contents of the flask are treated with 2 mL of
9.5% w/v potassium chloride solution and diluted to 100 mL
with 0.15% v/v nitric acid.

The barium content of the sample is determined by the method of standard additions. To 2-mL aliquots of the prepared
sample contained in separate 10-mL volumetric flasks, 5-mL
additions of 5-, 10-, and 15-mg/L barium standard are made.
No addition is made to a fourth flask containing a 2-mL
aliquot of the prepared sample. The contents of each flask
are treated with 0.06 mL of the 9.5% potassium chloride solution and diluted to 10 mL with 0.15% v/v nitric acid. The
spiked and unspiked samples are preferably aspirated into a
rich nitrous oxide-acetylene flame, and the absorbances are
measured at a wavelength of 553.6 nm.

For barium determinations by the furnace technique, method
7081, the sample is digested with nitric acid as described for
the flame technique. The colorless or pale yellow solution is
evaporated to near dryness, treated with 2 mL of 1:1 nitric
acid, and warmed to dissolve any residue. The watch glass and
inner walls of the beaker are washed down with high-purity
water, and the contents of the beaker are quantitatively
transferred to a 100-mL volumetric flask. Final dilution is
made with 0.15% v/v nitric acid.

The barium content of the sample is determined by the method of standard additions. To four separate 10-mL volumetric
flasks each containing 2 mL of prepared sample are made 5-mL
additions of 100-, 150-, and 200-µg/L barium standard, respectively. The fourth flask is the unspiked sample. The
contents of all four flasks are brought to volume with 0.15%
v/v nitric acid. The absorbances of replicate 20-µL injections of spiked and unspiked samples are measured at 553.6 nm
in nonpyrolytic graphite furnace tubes with continuous flow
purge gas using the following program: dry, 30 s at 125°C;
ash, 30 s at 1200°C; atomize, 10 s at 2800°C.

6.6.2. Canadian DOE Method 56050 for Barium

Method 56050 is applicable to the determination of barium
in sediments. Samples are prepared by the procedures described in Sections 6.2.5, 6.2.6, or 6.2.7. To compensate for
ionization interferences, both samples and standards are adjusted to 3000 ppm sodium. The samples and standards are

aspirated into a rich nitrous oxide-acetylene flame and their absorbances are measured at 553.6 nm. The barium content of the sample is determined by direct comparison. (The samples may be retained for subsequent determinations of calcium and strontium.)

The coefficients of variation at barium levels of 500 mg/ kg, 1000 mg/kg, and 1500 mg/kg were 11.9%, 10.6%, and 5.2%, respectively. Average recovery of spikes to a series of samples was 95%.

6.7. Procedures for Beryllium

The general procedures for the determination of beryllium are presented in Section 5.7.

6.7.1. Canadian DOE Method 04050 for Beryllium

Method 04050 is applicable to the determination of beryllium in sediments. Samples are prepared by the methods described in Sections 6.2.5, 6.2.6, and 6.2.7. Beryllium is vulnerable to interferences from aluminum, silicon, and other elements. Random samples should be split and spiked to determine whether these interferences are present. The method of standard additions may be employed to compensate for such interferences.

For the determination of beryllium by direct comparison, samples and standards are aspirated into a rich nitrous oxide-acetylene flame, and the absorbances are measured at 234.9 nm. The coefficients of variation at beryllium levels of 100 mg/ kg, 500 mg/kg, 1000 mg/kg, and 1500 mg/kg were 3.2%, 2.4%, 1.5%, and 1.5%, respectively. The recovery of spikes to a series of samples averaged 101%.

6.8. Procedures for Cadmium

Section 5.8 contains the general conditions for the determination of cadmium.

6.8.1. U.S. EPA Methods 7130 and 7131

Methods 7130 and 7131 contain approved procedures for the determination of cadmium in waste samples and in EP extracts by either flame or furnace atomic absorption spectrometry. The working range of the former is from 0.05 mg/L to 2 mg/L, and the latter is useful for the determination of cadmium in the concentration range of 0.5 µg/L to 10 µg/L.

Waste samples and EP extracts are prepared for aspiration or injection by the same nitric acid digestion used to prepare samples for the determination of barium (Section 6.6.1).

For the determination of cadmium by the flame technique, method 7130, the digested sample is evaporated to near dryness, treated with 5 mL of 1:1 hydrochloric acid, and warmed to dissolve any residue. This solution is cooled, quantitatively transferred to a 100-mL volumetric flask, and brought to volume with 0.15% v/v nitric acid.

The cadmium content of the samples is determined by the method of standard additions. Five-milliliter additions of 1.0-, 1.5-, and 2.0-mg/L cadmium standards are made to 2-mL aliquots of the prepared sample contained in 10-mL volumetric flasks. No addition is made to a fourth aliquot in a fourth flask. The contents of the flasks are brought to volume with 0.15% v/v nitric acid, and the spiked and unspiked samples are aspirated into an oxidizing air-acetylene flame. The absorbances are measured at 228.8 nm.

For the determination of cadmium by the furnace technique, method 7131, the digested sample is evaporated to near dryness, treated with 1 mL of 1:1 nitric acid, and warmed to dissolve any precipitate. This solution is quantitatively transferred to a 100-mL volumetric flask, treated with 2 mL of 40% w/v diammonium monohydrogen phosphate and brought to volume with 0.15% v/v nitric acid.

The cadmium content of the sample is determined by the method of standard additions. A 5-mL addition of 2.5-, 5.0-, or 10-µg/L cadmium standard is made to a 2-mL aliquot of the prepared sample contained in a 10-mL volumetric flask. No addition of a standard is made to the fourth 2-mL aliquot in a fourth flask. The contents of all four flasks are treated with 0.06 mL of the 40% w/v diammonium monohydrogen phosphate solution and brought to volume with high-purity water. The absorbances of replicate 20-µL injections of the spiked and unspiked samples are measured at 228.8 nm in nonpyrolytic graphite furnace tubes with continuous flow purge gas using the following program: dry, 30 s at 125°C; ash, 30 s at 500°C; atomize, 10 s at 1900°C.

6.8.2. Canadian DOE Method 48050 for Cadmium

This method is applicable to the determination of cadmium in sediments. Samples are prepared by the methods described in Sections 6.2.5, 6.2.6, or 6.2.7. Samples and standards are aspirated into a lean air-acetylene flame, and the absorbances are measured at 228.8 nm. The cadmium content of the samples is determined by direct comparison.

The coefficients of variation at cadmium levels of 100 mg/kg, 500 mg/kg, 1000 mg/kg, and 1500 mg/kg were 5.3%, 3.6%, 5.3%, and 4.3%, respectively. Recovery of cadmium spikes to a series of samples was 100%.

6.9. Procedures for Calcium

The general conditions for the determination of calcium are presented in Section 5.9.

6.9.1. Canadian DOE Method 20050 for Calcium

Method 20050 is applicable to the determination of calcium in sediments. Samples may be prepared by the procedures described in Sections 6.2.5, 6.2.6, or 6.2.7 (or samples previously prepared for the determination of barium, Section 6.6.2, may be used). Samples and standards are adjusted to 3000 ppm sodium to compensate for ionization interference.

Samples and standards are aspirated into a rich nitrous oxide-acetylene flame and the absorbances are measured at 422.7 nm. The calcium content of the samples is determined by direct comparison. The coefficient of variation at a calcium level of 10,000 mg/kg was 3.0%, and the recovery of calcium spikes from a series of samples averaged 97%.

6.10. Procedures for Chromium

The general conditions for the determination of chromium are presented in Section 5.10.

6.10.1. U.S. EPA Methods 7190 and 7191 for Chromium

Methods 7190 and 7191 contain the approved procedures for the preparation of EP extracts, industrial liquid wastes, land fill components or land fill leachates, and for the determination of chromium by flame or furnace atomic absorption spectrometry in these materials. The flame technique has an optimum working range of from 0.5 mg/L to 10 mg/L, and the furnace technique is best suited to the determination of chromium levels in the 5 µg/L to 100 µg/L range.

Samples or extracts are prepared for aspiration or injection by the same nitric acid digestion used to prepare samples or extracts for the determination of antimony (Section 6.4.1).

For the determination of chromium by flame atomic absorption spectrometry, method 7190, the digested sample is evaporated to near dryness, cooled, treated with 1 mL of nitric acid, and warmed to dissolve any precipitated material. The contents of the beaker are quantitatively transferred to a 100-mL volumetric flask. If the air-acetylene flame is used, the contents of the flask are treated with 1 mL of a solution containing 200 mg sodium sulfate and 1 g ammonium bifluoride per 100 mL. The sample solution is brought to volume with high-purity water.

The chromium content of the sample is determined by the method of standard additions. To three 2-mL aliquots of the sample solution contained in 10-mL volumetric flasks, 5-mL additions of 5-, 8-, and 10-mg/L chromium standard are made, respectively. No addition is made to the fourth flask, which also contains a 2-mL aliquot of the sample solution. The contents of all four flasks are brought to volume with 0.15% v/v nitric acid. The absorbances are measured at 357.9 nm in a rich nitrous oxide-acetylene flame.

For the determination of chromium by the furnace technique, method 7191, the digested sample is evaporated to near dryness, cooled, treated with 1 mL of 1:1 nitric acid, and warmed to dissolve any precipitated material. The contents of the beaker are cooled and treated with 1 mL of 30% hydrogen peroxide and 1 mL of a solution made by dissolving 11.8 g of calcium nitrate tetrahydrate in 100 mL of high-purity water. The contents of the beaker are quantitatively transferred to a 100 mL volumetric flask and brought to volume with high-purity water.

The chromium contents of the sample are determined by the method of standard addition. Two-milliliter aliquots of the sample solution are transferred to four separate 10-mL volumetric flasks. No addition is made to the first flask, and 5-mL additions of 50-, 75-, and 100-µg/L chromium standards are made to the remaining flasks, respectively. The contents of the 10-mL flasks are treated with 0.05 mL of 30% hydrogen peroxide and 0.05 mL of the calcium nitrate solution cited above. The contents of the flasks are brought to volume with high-purity water. The absorbances of replicate 20-μL injections are measured at 357.9 nm in nonpyrolytic graphite furnace tubes with continuous flow purge gas. The following furnace program is recommended: dry, 30 s at 125°C; ash, 30 s at 1000°C; atomize, 10 s at 2700°C.

6.10.2. Canadian DOE Method 24050 for Chromium

Method 24050 is applicable to the determination of chromium in sediments. The samples are prepared by the procedures described in Sections 6.2.5, 6.2.6, or 6.2.7. Possible interferences from iron can be controlled by making sample and standard solutions 2% w/v in ammonium chloride. The samples and standards are aspirated into a reducing air-acetylene flame, and the absorbances are measured at 357.9 nm. The chromium content of the samples is determined by direct comparison.

The coefficient of variation at chromium levels of 100 mg/kg, 500 mg/kg, 1000 mg/kg, and 1500 mg/kg were 7.7%, 5.7%, 2.9%, and 3.0%, respectively. The average recovery of spikes to a series of samples was 99%.

6.11. Procedures for Cobalt

Section 5.11 contains the general conditions for the determination of cobalt by atomic absorption spectrometry.

6.11.1. Canadian DOE Method 27050 for Cobalt

Method 27050 is applicable to the determination of cobalt in sediments. Samples are prepared by the procedures described in Sections 6.2.5, 6.2.6, or 6.2.7. Samples and standards are aspirated into a lean air-acetylene flame, and the absorbances are measured at 240.7 nm. The cobalt content of the samples is determined by direct comparison.

The coefficients of variation at cobalt levels of 100 mg/kg, 500 mg/kg, 1000 mg/kg, and 1500 mg/kg were 10.5%, 10.7%, 5.1%, and 5.4%, respectively. Recoveries of cobalt spikes to a series of samples averaged 97%.

6.12. Procedures for Copper

The general conditions for the determination of copper by atomic absorption spectrometry are presented in Section 5.12.

6.12.1. Canadian DOE Method 29050 for Copper

Method 29050 is applicable to the determination of copper in sediments. The samples are prepared by the procedures described in Sections 6.2.5, 6.2.6, or 6.2.7. Samples and standards are aspirated into a lean air-acetylene flame, and the absorbances are measured at 324.7 nm. The copper content of the samples is determined by direct comparison.

The coefficients of variation at copper levels of 100 mg/kg, 500 mg/kg, 1000 mg/kg, and 1500 mg/kg were 2.0%, 1.6%, 1.5%, and 1.6%, respectively. The recovery of copper spikes to a series of samples averaged 95%.

6.13. Procedures for Iron

The general conditions for the determination of iron are presented in Section 5.15.

6.13.1. Canadian DOE Method 26050 for Iron

This method is applicable to the determination of iron in sediments. Samples are prepared by the procedures described in Sections 6.2.5, 6.2.6, or 6.2.7. Because nitric acid suppresses the absorbance of iron, the acid matrix of the samples and standards is matched as closely as possible. Samples and standards are aspirated into a lean air-acetylene flame, and absorbances at 248.3 nm are recorded. The iron content of the samples is determined by direct comparison.

The coefficients of variation at iron levels of 5000 mg/kg, 10,000 mg/kg, and 15,000 mg/kg were 9.6%, 6.3%, and 7.0%, respectively. The average recovery of iron spikes added to a series of samples was 102%.

6.14. Procedures for Lead

Section 5.16 contains the general conditions for the determination of lead.

6.14.1. U.S. EPA Methods 7420 and 7421 for Lead

Methods 7420 and 7421 contain the approved procedures for the determination of lead in waste samples and in EP extracts. The samples or extracts are prepared by nitric acid digestion, and their lead content is determined by either flame or furnace atomic absorption spectrometry. The flame technique is applicable to lead concentrations in the 1 mg/L to 20 mg/L range, and the furnace technique is best suited to lead concentrations of from 5 µg/L to 100 µg/L.

Waste samples or EP extracts are prepared for aspiration or injection by the same nitric acid digestion as that employed prior to the determination of antimony (Section 6.4.1).

For the determination of lead by the flame technique, method 7420, digested sample is evaporated to near dryness, treated with 5 mL of 1:1 hydrochloric acid, warmed to dissolve precipitated material, and quantitatively transferred to a 100-mL volumetric flask. The contents of the flask are brought to volume with high-purity water, and four 2-mL aliquots are transferred to separate 10-mL volumetric flasks for the determination of lead by the standard additions method. No addition is made to the first flask, and 5-mL additions of 10-, 15-, and 20-mg/L lead standards are made to the remaining flasks, respectively. The spiked and unspiked samples are aspirated into a lean air-acetylene flame, and their absorbances are recorded at 283.3 nm.

For the furnace technique, method 7421, the digested sample is evaporated to near dryness, treated with 1 mL of 1:1 nitric acid, heated to dissolve precipitated material, and cooled. The solution is transferred to a 100-mL volumetric flask, treated with 10 mL of lanthanum nitrate solution (58.64 g of lanthanum oxide dissolved in 100 mL of concentrated nitric acid and then diluted to 1 L with high-purity water), and brought to volume with high-purity water.

The lead content of the samples is determined by the method of standard additions. Four 2-mL aliquots of the sample solution are pipetted into separate 10-mL volumetric flasks. No addition is made to the first flask. Additions of 5 mL of 50-, 75-, and 100-µg/L lead standards are made to the remaining flasks, respectively. The absorbances of replicate 20-µL injections into nonpyrolytic graphite furnace tubes are measured

at 283.3 nm with continuous purge gas flow. The following
furnace program is recommended: dry, 30 s at 125°C; ash, 30 s
at 500°C; atomize, 10 s at 2700°c.

6.14.2. Canadian DOE Method 82050 for Lead

Method 82050 is applicable to the determination of lead in
sediments. Samples are prepared by the procedures described
in Sections 6.2.5, 6.2.6, or 6.2.7. The samples and standards
are aspirated into a lean air-acetylene flame, and the absorbances are measured at 283.3 nm. The lead content of the
samples is determined by direct comparison.

The coefficient of variation at lead levels of 100 mg/kg,
500 mg/kg, 1000 mg/kg, and 1500 mg/kg were 10.0%, 8.2%, 7.2%,
and 4.8%, respectively. The average recovery of lead spikes
added to a series of samples was 97%.

6.15. Procedures for Lithium

The general conditions for the determination of lithium
are presented in Section 5.17.

6.15.1. Canadian DOE Method 03050 for Lithium

Method 03050 is applicable to the determination of lithium
in sediments. Samples are prepared by the procedures described
in Sections 6.2.5, 6.2.6, or 6.2.7. Sample and standards are
aspirated into a lean air-acetylene flame, and their absorbances
are measured at 670.8 nm. The lithium content of the samples is
determined by direct comparison.

The coefficients of variation at lithium levels of 100 mg/kg,
500 mg/kg, 1000 mg/kg, and 1500 mg/kg were 3.5%, 3.2%, 0.9%,
and 1.5%, respectively. The average recoveries of lithium
spikes added to a series of samples was 99%.

6.16. Procedures for Magnesium

The general conditions for the determination of magnesium
are presented in Section 5.18.

6.16.1. Canadian DOE Method 12050 for Magnesium

Method 12050 is applicable to the determination of magnesium in sediments. The samples are prepared by the methods described in Sections 6.2.5, 6.2.6, or 6.2.7. Samples and standards are aspirated into a rich nitrous oxide-acetylene flame, and their absorbances at 285.2 nm are recorded. The magnesium content of the samples is determined by direct comparison.

At a magnesium level of 10,000 mg/kg, the coefficient of variation was 2.9%. The average recovery of magnesium spikes to a series of samples was 98%.

6.17. Procedures for Manganese

Section 5.19 contains a description of the general conditions for the determination of manganese.

6.17.1. Canadian DOE Method 25050 for Manganese

Method 25050 is applicable to the determination of manganese in sediments. Samples are prepared by the procedures described in Sections 6.2.5, 6.2.6, or 6.2.7. The hydrofluoric acid and the boric acid from the latter procedure depresses the sensitivity for the determination of manganese. Matrix matching is necessary to overcome this depression. Samples and standards are aspirated into a lean air-acetylene flame, and their absorbances at 279.5 nm are recorded. The manganese content of the samples is determined by direct comparison.

The coefficients of variation at manganese levels of 500 mg/kg, 1000 mg/kg, and 1500 mg/kg were 4.5%, 1.5%, and 3.2%, respectively. A standard addition to a series of samples produced an average recovery of 98%.

6.18. Procedures for Mercury

The general conditions for the determination of mercury are presented in Section 5.20. The procedures for the determination of mercury in liquid wastes, solid wastes, sludges, sediments, and soils utilize the cold vapor technique. The loss of mercury during sample preparation and the volatilization

of ultraviolet-absorbing organic or inorganic materials while
making absorbance measurements are potential sources of error
requiring special consideration.

6.18.1. U.S. EPA Method 245.5 for Mercury

Method 245.5 is applicable to the determination of total
(inorganic and organic) mercury in soils, sediments, bottom
deposits, and sludge-type materials. Samples may be oven
dried at 60°C without mercury loss. The dried material is
pulverized and mixed thoroughly, and triplicate 0.2-g subsamples are weighed into biochemical oxygen demand (BOD) bottles.* The contents of the bottles are treated with 5 mL of
aqua regia and heated for 2 min in a 95°C water bath. The
bottles are cooled, treated with 50 mL of high-purity water
and 15 mL of 5% w/v potassium permanganate solution, and returned to the water bath for 30 min. The samples are cooled,
brought to a volume of 100 mL with high-purity water, and
treated with 6 mL of 12% w/v sodium chloride-12% w/v hydroxylamine sulfate solution. The bottles are individually treated
with 5 mL of 10% w/v stannous sulfate suspension in 0.5 N
sulfuric acid and immediately connected to the aeration apparatus. The mercury content of the samples is determined by
direct comparison using the maximum absorbances at 253.7 nm
which is usually exhibited within 30 s when the circulating
pump is operated at 1 L/min.

The following results were obtained by replicate mercury
determination on two sediment samples:

---

*As an alternate procedure, the contents of the bottles may be
  treated with 5 mL of sulfuric acid, 2 mL of nitric acid, and
  5 mL of 5% w/v potassium permanganate solution. The bottles
  are closed with aluminum foil and autoclaved for 15 min at
  121°C and 5 lb. The autoclaved samples are cooled, brought
  to a volume of 100 mL with high-purity water, and treated
  with 6 mL of 12% w/v sodium chloride-12% w/v hydroxylamine
  sulfate solution.

	Sample 1	Sample 2
Mean (mg Hg/kg)	0.29	0.82
Coefficient of variation (%)	6.9	3.7

The recoveries of methyl mercuric chloride spikes to these samples were 97% and 94%, respectively.

6.18.2. U.S. EPA Method 7470 for Mercury

Method 7470 contains the approved procedures for the determination of mercury in an EP extract, in a solid or liquid waste, or in a landfill leachate. Four 10-mL aliquots of each sample are transferred to separate 50-mL volumetric flasks. No addition is made to the first flask. Additions of 25-mL of mercury standards, 50 µg/L, 80 µg/L, and 100 µg/L, are made to the second, third, and fourth flasks, respectively. The contents of the flasks are brought to volume with 0.15% v/v nitric acid. Ten-milliliter aliquots are transferred from each volumetric flask to separate BOD bottles, and they are diluted to 100 mL with high-purity water. The contents of each bottle are treated with 5 mL of 0.5 N sulfuric acid, 2.5 mL of concentrated nitric acid, and 15 mL of 5% w/v potassium permanganate solution. If the color of the permanganate is discharged within 15 min, additional 15-mL increments are added until the color persists for this period of time. Eight milliliters of 5% w/v potassium persulfate solution are added to each bottle, and the bottles are heated in a 95°C water bath for 2 h. After this time, the contents of each bottle are treated with 6 mL of 12% w/v sodium chloride–12% w/v hydroxylamine sulfate solution, and the head space of each bottle is purged with compressed air 2 min later. The bottles are _individually_ treated with 5 mL of 10% w/v stannous sulfate suspension in 0.5 N sulfuric acid and _immediately_ connected to the aeration apparatus. The absorbances are measured at 253.7 nm. The maximum value is usually observed within 30 s when the circulating pump is operating at 1 L/min. The method of standard additions, using the maximum absorbances, is employed to determine the mercury content of the samples.

6.18.3. Canadian DOE Method 80050 for Mercury

Method 80050 is applicable to the determination of mercury in soils and sediments. Samples are collected with an Ekman dredge, a grab sampler, or a core sampler depending on conditions and requirements. The samples should be drained immediately after collection and stored in polypropylene bags or bottles under refrigeration. A representative portion of the wet sample equivalent to 1-2 g dry weight* is weighed into a 100-mL volumetric flask; cooled in an ice bath; treated with 10 mL of sulfuric acid, 5 mL of nitric acid, and 2 mL of hydrochloric acid; and incubated in a shaking water bath for 2 h at 50° to 60°C. After this time, the contents of the flasks are cooled and cautiously treated with 15 mL of 6% w/v potassium permanganate solution in 1 mL increments; 30 min later, 5 mL of 5% potassium persulfate solution are added. The flasks are allowed to stand overnight. If the purple color of the permanganate fades, more permanganate is added until the purple color persists for 15 min. The contents of the flasks are treated with 10 mL of 6% w/v hydroxylamine sulfate-6% w/v sodium chloride solution and brought to volume with high-purity water. The mercury content of the sample is determined by direct comparison using the automated cold vapor technique shown in Figure 6.6.

At mercury levels of 0.1 mg/kg, 0.6 mg/kg, 1 mg/kg, and 1.6 mg/kg, the coefficients of variation for replicate determinations were 10%, 4%, 2%, and 2%, respectively. Soil samples spiked with 0.1 µg and with 0.2 µg of mercury from mercuric chloride showed recoveries of from 97% to 107%. Sample spiked at similar levels with phenyl mercuric acetate showed similar recoveries.

6.19. Procedures for Molybdenum

The general conditions for the determination of molybdenum are presented in Section 5.21.

---

*The moisture content is determined by measuring the loss in weight shown by a wet sample after drying for 24 h at 105°C.

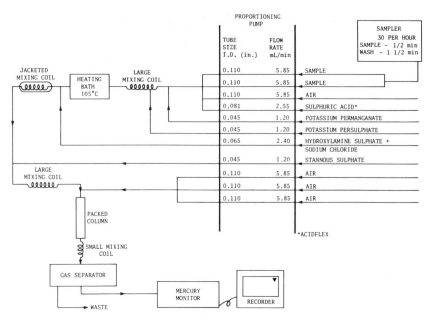

Figure 6.6. Mercury manifold.[13]

6.19.1. Canadian DOE Method 82050 for Molybdenum

Method 82050 is applicable to the determination of molybdenum in sediments. The samples are collected and stored in accord with Section 6.1.2, and they are prepared by the procedures described in Sections 6.2.5, 6.2.6, or 6.2.7. Ten-
milliliter aliquots of the standard solutions and of the
prepared samples are pipetted into plastic tubes, treated with
0.2 mL of 2.5% w/v aluminum chloride solution, and aspirated
into a rich nitrous oxide-acetylene flame. Absorbances at
313.3 nm are recorded, and the molybdenum content of the
samples is determined by direct comparison. The samples in
the plastic tubes may be retained for the determination of
vanadium.

The coefficients of variation at molybdenum levels of
100 mg/kg, 500 mg/kg, 1000 mg/kg, and 1500 mg/kg were 4.0%,
3.2%, 2.4%, and 1.1%, respectively. The recovery of molybdenum spikes to a series of samples averaged 100%.

6.20. Procedures for Nickel

Section 5.22 contains the general conditions for the determination of nickel.

6.20.1. U.S. EPA Methods 7520 and 7521 for Nickel

Methods 7520 and 7521 describe the approved procedures for the premeasurement preparation of wastes and EP extracts and for the measurement of their nickel contents by flame or furnace atomic absorption spectrometry.

A 100-g waste sample or 100 mL of EP extract is transferred to a 250-mL beaker, treated with 3 mL of concentrated nitric acid, and digested by the procedures described in Section 6.4.1. The digested material is treated with 5 mL of 1:1 nitric acid, warmed to dissolve any precipitated material, cooled, and diluted to 100 mL with high-purity water. The nickel content of the sample is determined by the method of standard additions. The flame technique is most useful for solutions containing from 0.3 mg/L to 5 mg/L, while the graphite furnace is suitable for the 5 µg/L to 100 µg/L concentration range.

For the determination of nickel by flame atomic absorption spectrometry, method 7520, 2-mL aliquots of the prepared sample are pipetted into 10-mL volumetric flasks, and 5-mL additions of 1-, 3-, and 5- mg/L nickel standards are made to the second, third, and fourth flasks, respectively. The contents of all four flasks are brought to volume with 0.5% v/v nitric acid, and their absorbances are measured at 232.0 nm (352.4 nm, which is less susceptible to interference but of diminished sensitivity, may also be used) in a lean air-acetylene flame.

For the determination of nickel by the furnace technique, method 7521, 2-mL aliquots of the prepared sample are pipetted into 10-mL volumetric flasks, and 5-mL additions of 25 µg/L, 50 µg/L, and 75 µg/L of nickel standards are made to the second, third, and fourth flasks, respectively. The contents of all four flasks are brought to volume with 0.5% v/v nitric acid, and the absorbances of replicate 20-µL injections are measured at 232.0 nm in nonpyrolytic graphite tubes with

continuous gas flow using the following program: dry, 30 s at
125°C; ash, 30 s at 900°C; atomize, 10 s at 2700°C.

6.20.2. Canadian DOE Method 28050 for Nickel

This method is applicable to the determination of nickel in
sediments. Samples are collected and stored in accord with
Section 6.1.2, and they are prepared for atomic absorption
spectrometry by the procedures described in Sections 6.2.5,
6.2.6, or 6.2.7. Standards and prepared samples are aspirated
into a lean air-acetylene flame, and their absorbances are
measured at 232.0 nm. The nickel content of the sample is
determined by direct comparison.

The coefficients of variation at nickel levels of 100 mg/
kg, 500 mg/kg, 1000 mg/kg, and 1500 mg/kg were 10.1%, 9.4%,
5.8%, and 12.6%, respectively. Recovery of spikes to a series
of samples averaged 98%.

6.21. Procedures for Selenium

The general conditions for the determination of selenium
are presented in Section 5.30. Both the hydride technique and
the furnace technique are suitable for the determination of
selenium in wastes, sludges, sediments, and soils.

6.21.1. U.S. EPA Methods 7740 and 7741 for Selenium

Methods 7740 and 7741 contain the approved procedures for
the determination of selenium in wastes and in EP extracts.
Sample preparation and subsequent absorption measurements by
furnace or hydride techniques are described.

To prepare samples for the determination of selenium by the
furnace technique, method 7740, a 100-g portion of the waste
or 100 mL of the EP extract is transferred to a 250-mL beaker,
treated with 1 mL of concentrated nitric acid and 2 mL of 30%
hydrogen peroxide, and heated on a hotplate at 95°C for 1 h or
until the volume is reduced to less than 50 mL. Additional
nitric acid and hydrogen peroxide may be required to complete
the digestion. Such additions should be made until no further
changes are noted in the amount of particulate matter remaining. The digestion mixture is cooled, diluted to 50 mL with

high-purity water, and centrifuged. A 25-mL aliquot of the
supernatant is transferred to a 50-mL volumetric flask, treated with 5 mL of 1% w/v nickel nitrate solution, and brought to
volume with high-purity water. This solution is used to determine the selenium contents of the sample with the furnace
technique by the method of standard additions.

Four 10-mL volumetric flasks, each containing 2 mL of the
prepared sample, are treated with 5 mL of either high-purity
water, 50-µg/L selenium standard, 75-µg/L selenium standard,
or 100-µg/L selenium standard, respectively. The contents of
the flasks are then treated with 0.3 mL of 1% w/v nickel
nitrate solution and 0.3 mL of 30% hydrogen peroxide and
brought to volume with high-purity water. The absorbances of
replicate 20-µL injections are measured at 196.0 nm in nonpyrolytic graphite tubes with continuous flow purge gas using
the following furnace program: dry, 30 s at 125°C; ash, 30 s
at 1200°C; atomize, 10 s at 2700°C.

For the hydride technique, method 7741, a 50-g waste sample or 50 mL of EP extract is transferred to a 250-mL beaker,
treated with 10 mL of concentrated nitric acid and 12 mL of
18 N sulfuric acid, and heated on a hotplate until dense white
fumes of SO_3 are evolved. It is necessary to prevent charring
of the sample during the digestion process by maintaining an
excess of nitric acid. When the sample remains colorless (or
pale straw) while evolving SO_3 fumes, digestion is complete.
The sample is cooled, diluted to 50 mL with high-purity water,
and again heated until dense white fumes of SO_3 are evolved.
After it has cooled to room temperature, the sample is transferred to a 100-mL volumetric flask, treated with 40 mL of
concentrated hydrochloric acid, and brought to volume with
high-purity water.

Ten-milliliter aliquots of the prepared sample solution are
transferred to separate 50-mL volumetric flasks. No addition
is made to the first flask, and 25 mL of selenium standards
containing 15 µg/L, 20 µg/L, or 25 µg/L are added to the second, third, and fourth flasks, respectively. The contents of
the flasks are brought to volume with 0.15% v/v nitric acid,
and 25-mL aliquots of the sample or spiked sample are pipetted
into the reaction vessel of the hydride generator. The contents of the reaction vessel are treated with 0.5 mL of stannous chloride solution (100 g dissolved in 100 mL of concentrated hydrochloric acid) and allowed to stand for 10 min.

The reaction vessels are individually connected to the hydride generator and injected with 1.5 mL of zinc slurry. The absorbances at 196.0 nm are measured in the argon-hydrogen flame, and the selenium content of the sample is determined by the method of standard additions.

6.22. Procedures for Silver

The general conditions for the determination of silver are presented in Section 5.31.

6.22.1. U.S. EPA Methods 7760 and 7761 for Silver

Methods 7760 and 7761 have been approved for the determination of silver in wastes and in EP extracts. A 100-mL sample, contained in a 250-mL beaker, is treated with 3 mL of concentrated nitric acid and heated to near dryness on a hotplate. Another 3 mL of nitric acid are added, the beaker is covered with a watch glass, and the contents are heated at a gentle reflux. Additional 3-mL increments of nitric acid are added and the refluxing is continued until the digestion is complete as indicated by a light colored solution. The solution is evaporated to near dryness, treated with 2 mL of 1:1 nitric acid, and allowed to cool. The solution is then rendered ammoniacal, treated with 1 mL of cyanogen iodide solution,* and allowed to stand for 1 h. This solution is quantitatively transferred to a 100 mL volumetric flask and brought to volume with high-purity water. The silver content of the prepared sample is determined by the method of standard additions using either flame or furnace techniques.

For the flame technique, method 7760, 2-mL aliquots of the prepared sample are pipetted into each of four 10-mL volumetric flasks, and 5 mL of high-purity water are added to the first flask, 5 mL of 2-mg/L silver standard to the second,

---

*Cyanogen iodide solution is prepared by dissolving 4.0 mL of concentrated ammonium hydroxide, 6.5 g of potassium cyanide, and 5.0 mL of 1.0 N iodine solution in 50 mL of high-purity water and diluting to a final volume of 100 mL.

5 mL of 3-mg/L silver standard to the third, and 5 mL of
4-mg/L silver standard to the fourth flask. The contents of
the flasks are brought to volume with a 1:100 dilution of the
cyanogen iodide solution. The sample and the spiked samples
are aspirated into a lean air-acetylene flame, and the absorbances at 328.1 nm are recorded.

For the furnace technique, method 7761, three of four 2-mL
aliquots of the prepared sample contained in separate 10-mL
volumetric flasks are spiked with 5 mL of 5 µg/L, 10 µg/L, and
15 µg/L silver standards, respectively. All four flasks are
brought to volume with a 1:100 dilution of the cyanogen iodide
solution, and the absorbances of replicate 20 µL injections
are measured at 328.1 nm in nonpyrolytic graphite tubes with
continuous flow argon purge gas using the following program:
dry, 30 s at 125°C; ash, 30 s at 400°C; atomize, 10 s at
2700°C.

6.23. Procedures for Vanadium

The general conditions for the determination of vanadium
are presented in Section 5.37.

6.23.1. Canadian DOE Method 23050 for Vanadium

Method 23050 is applicable to the determination of vanadium
in sediments. The samples are collected and stored in accord
with Section 6.1.2 and prepared for the determination of vanadium by the procedures described in Sections 6.2.5, 6.2.6, or
6.2.7. Ten-milliliter aliquots of the standard solutions and
the prepared sample solutions are pipetted into plastic tubes,
treated with 0.2 mL of 2.5% w/v aluminum chloride solution,
and aspirated into a rich nitrous oxide-acetylene flame. Absorbances at 318.4 are recorded, and the vanadium content of
the samples is determined by direct comparison. The solutions
may be retained for the determination of molybdenum.

The coefficients of variation at vanadium levels of 500 mg/
kg, 1000 mg/kg, and 1500 mg/kg were 9.6%, 5.0%, and 5.3%, respectively. The recovery of spikes added to a series of samples averaged 103%.

6.24. Procedures for Zinc

The general conditions for the determination of zinc are presented in Section 5.38.

6.24.1. Canadian DOE Method 30050 for Zinc

Method 30050 is applicable to the determination of zinc in sediments. Samples are collected and stored in accord with Section 6.1.2, and they are prepared for the determination of zinc by the procedures described in Sections 6.2.5, 6.2.6, or 6.2.7. The prepared samples and standards are aspirated into a lean air-acetylene flame, and their absorbances at 213.9 nm are recorded. The zinc content of the samples is determined by direct comparison.

The coefficients of variation at zinc levels of 500 mg/kg, 1000 mg/kg, and 1500 mg/kg were 7.4%, 5.5%, and 2.9%, respectively. The recovery of spikes added to a series of samples averaged 96%.

7. METHODS FOR MONITORING TRACE METALS IN ANIMAL TISSUES AND BODY FLUIDS

The Association of Official Analytical Chemists (AOAC) Official Methods of Analysis[164] and the U.S. Federal Drug Administration (FDA) Laboratory Information Bulletins[29] contain procedures for the determination of metals in foods by atomic absorption spectrometry, and this technique serves as the basis for the determination of trace metals in blood and urine according to several manuals on clinical chemistry.[157-169] Procedures for monitoring the occupational and natural environments are found in the National Institute of Occupational Safery and Health (NIOSH) Manual[10] and in the Canadian (DOE) Analytical Methods Manual.[13] The former contains the regulatory compliance methods for monitoring toxic metal levels in the blood and urine of occupationally exposed humans pursuant to the federal (U.S.) regulations. The Canadian DOE Manual describes procedures for the determination of metals in fish tissues.

7.1. Canadian DOE Method for Arsenic and Selenium in Fish

For the determination of arsenic and selenium in fish tissue, the sample is homogenized and a weighed aliquot is digested with nitric-perchloric-sulfuric acid mixture. The arsenic and selenium contents of the digested material are determined sequentially by automated, flameless atomic absorption spectrometry of the respective hydrides. Recoveries of inorganic and organic spikes carried through the procedure were from 90% to 102% for both arsenic and selenium, and the determination of arsenic and selenium in bovine liver [U.S. National Bureau of Standards (NBS) standard reference material (SRM) No. 1577] and in orchard leaves (U.S. NBS SRM No. 1571) showed excellent agreement with the certified values. The precision of the method is reflected in the following table:

	Arsenic		Selenium	
Sample	Mean (mg/kg)	CV (%)	Mean (mg/kg)	CV (%)
Smelt 1	0.444	15	0.316	6
Smelt 2	0.412	5	0.308	5
Salmon 1	0.260	7	0.376	6
Salmon 2	0.366	7	0.548	5

The fish are stored frozen; they are thawed at room temperature prior to homogenization and sampling. Individual
fish, specific fish organs or tissues, or several fish (for
compositing) are homogenized by recycling the material through
a meat grinder several times or by making a puree in a tissue
homogenizer.*

A 1-g sample of the homogenized material is weighed into a
calibrated 50-mL test tube, treated with 25 mL of nitric acid,
and allowed to stand overnight. On the following day, the
contents of the test tube are treated with 5 mL of sulfuric
acid and 5 mL of perchloric acid. A boiling chip is added,
and the test tube is transferred to an aluminum heating block
maintained at 140°C. When the volume of solution is reduced
to 25 mL, the temperature of the heating block is raised to
175°C. The heating is discontinued and 5 mL more of nitric
acid are added if there is any indication of charring. The
heating is continued until the volume of solution is reduced
to 5 mL and dense white fumes of perchloric acid are no longer evolved. At this point, the test tube is removed from the
heating block and allowed to cool. The digested sample should
be colorless. It is diluted to 50 mL with high-purity water
and transferred to the autosampler.

Arsenic and selenium are determined sequentially as the
gaseous hydrides using the automated reduction-sparging manifolds shown in Figures 7.1 and 7.2, respectively. The conditions for operating these manifolds are as follows:

---

*Polytron Homogenizer, Brinkman Instruments.

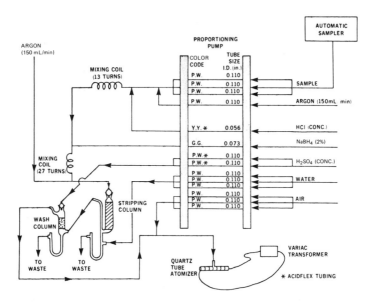

Figure 7.1. Arsenic manifold. *Acidflex tubing; P, purple; W, white; Y, yellow; G, green.

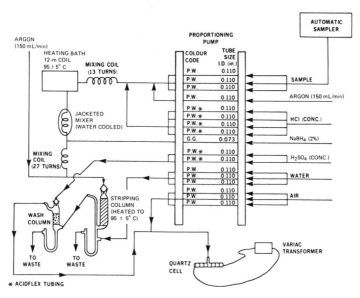

Figure 7.2. Selenium manifold. *Acidflex tubing; P, purple; W, white; Y, yellow; G, green.

	Arsenic	Selenium
Heating bath temperature (°C)	20-95	95 ± 5
Tube furnace temperature (°C)	800	800
Stripping column temperature (°C)	20-95	95 ± 5
Argon flow (mL/min)	150	150
Wavelength (nm)	193.7	196.0

The determination is made by direct comparison.

7.2. NIOSH Method P&CAM 279 for Beryllium in Biological Tissue

Method P&CAM 279 is applicable to the determination of beryllium in biological tissues for regulatory compliance monitoring under the U.S. Occupational Safety and Health Act (OSHA). Biopsy or autopsy specimens are digested with nitric-perchloric-sulfuric acid mixture, and the beryllium content of the resulting solution is measured by furnace atomic absorption spectrometry at 234.9 nm. The working range of the method is from 2 µg/kg to 250 µg/kg. The calcium interference is masked by use of a 3% sulfuric acid matrix. Beryllium recovery from spiked and unspiked samples of bovine liver (U.S. NBS SRM No. 1577) was 100%. Replicate determinations showed a relative precision of 10%.

Biological tissue samples are obtained from biopsy or autopsy specimens. These are stored frozen in plastic containers.

A 500-mg sample is weighed into a 125-mL beaker and treated with 3 mL of nitric acid, 1 mL of perchloric acid, and 1 mL of sulfuric acid. The beaker is covered with a watch glass, and its contents are heated at 130° to 150°C on a hotplate. The heating is continued, with additional incrementing of acid as needed, until a clear solution is obtained. The watch glass is then removed from the beaker, and the solution is allowed to evaporate to near dryness. The cooled residue is dissolved in 5 mL of 3% v/v sulfuric acid. The sample solution is quantitatively transferred to a 10-mL volumetric flask and brought to volume with 3% v/v sulfuric acid.

The absorbances of replicate 10-µL injections of samples and standards are measured at 243.9 nm using the following program: dry, 20 s at 110°C; ash, 25 s at 500°C; atomize, 8 s at 2800°C. The beryllium content of the samples is determined by direct comparison.

7.3. Canadian DOE Method for Mercury in Fish

For the determination of mercury in fish tissue, the sample is homogenized and a weighed aliquot of the homogenizate is subjected to low-temperature, 60°C, decomposition with either hydrogen peroxide and sulfuric acid or nitric-sulfuric acid mixture. The organomercurials in the decomposed material are oxidized to inorganic mercury compounds with permanganate and persulfate. After reduction of excess oxidants with hydroxy-lamine, the mercury content of the prepared sample is deter-mined by an automated cold vapor technique. The recovery of methyl mercuric chloride spikes to fish samples was 95% to 105% by this procedure, and the coefficients of variation at mercury levels of 0.14 mg/kg, 0.30 mg/kg, and 0.49 mg/kg were 9%, 5%, and 6%, respectively. Excellent agreement between certified and experimental values for the mercury content of orchard leaves (U.S. NBS SRM No. 1571) was obtained.

The fish samples are stored and homogenized in accord with the procedures described in Section 7.1. The homogenized ma-terial is digested by either of the following procedures:

7.3.1. Procedure 1

A 1-g aliquot of the homogenized material is weighed into a 100-mL volumetric flask, treated with 1 mL of 30% hydrogen peroxide, and allowed to stand for 10 min. After this time, 10 mL of sulfuric acid are added slowly while the flask is cooling in an ice bath. The flask is allowed to stand for 5 min at room temperature. If solubilization is not complete in this time, the flask is placed in a shaking water bath at 60°C for 30 min. The flask is returned to the ice bath and 20 mL of 5% w/v potassium permanganate solution are added slowly. After standing for 30 min, 10 mL of 5% w/v potassium persulfate solution are added. At this point, the contents of the flask are allowed to stand at room temperature overnight. The purple color of the permanganate should be evident on the following day. If this is not the case, more permanganate

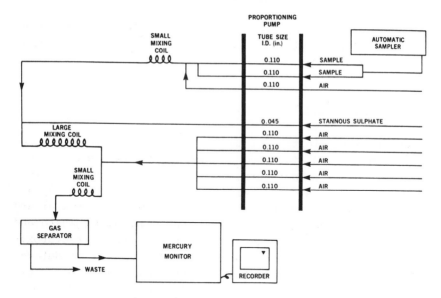

Figure 7.3. Mercury manifold.

solution is added in 20-mL increments until the purple color
persists for 15 min. The contents of the flask are then treated with 5 mL of 20% w/v hydroxylamine sulfate-20% w/v sodium
chloride solution and brought to volume with high-purity
water.

7.3.2. Procedure 2

A 1-g aliquot of the fish homogenizate is weighed into a
100-mL volumetric flask. The contents of the flask are treated with 10 mL of sulfuric acid and 5 mL of nitric acid, and
the flask is placed into a 60°C shaking water bath for 2 h.
The flask is then cooled in an ice bath, and its contents are
treated with 10 mL of 5% w/v potassium permanganate solution
followed by 10 mL of 5% w/v potassium persulfate solution
30 min later. The flask is then allowed to stand at room temperature overnight. If, on the following day, the purple col-
or of the permanganate is not observed, further additions are
made until the color persists for 15 min. The contents of the
flask are treated with 5 mL of 20% w/v hydroxylamine sulfate-
20% w/v sodium chloride solution and brought to volume with
high-purity water.

The mercury absorbances are measured by an automated cold vapor technique the manifold for which is shown in Figure 7.3. The mercury content of the sample is determined by direct comparison.

7.4. Canadian DOE Method for Trace Metals in Fish

For the determination of cadmium, chromium, copper, nickel, lead, and zinc in fish tissue, the sample is homogenized and a weighed aliquot is digested with nitric-perchloric-sulfuric acid mixture. The digestion is completed with hydrogen peroxide, and the levels of zinc, copper, and chromium in the resulting solution are determined by conventional flame atomic absorption spectrometry. The levels of nickel, lead, and cadmium are determined after concentration by the ammonium pyrrolidine thiocarbamate-methyl isobutyl ketone (APDC-MIBK) extraction procedure (Section 5.2.2). The accuracy and precision of the method have been evaluated with several standard reference materials, and statistical data from actual analyses of contaminated fish are available. This information is presented in Tables 7.1 and 7.2, respectively.

Fish samples are stored and homogenized under the same conditions as those described in Section 7.1. A 5-g aliquot of the homogenized material is weighed into a calibrated 50-mL test tube and treated with 5 mL of nitric acid and 5 mL of sulfuric acid. The mixture is allowed to stand until any initial reaction subsides. The test tube is then placed in an aluminum heating block and heated at 60°C for 30 min. The tube is removed from the heating block, allowed to cool for 5 min, and then treated with an additional 10 mL of nitric acid. The tube is returned to the heating block, and its contents are heated at increasing temperatures until 120°C is attained. Heating at this temperature is continued until the volume in the tube is reduced to 15 mL. The temperature is increased to 150°C, and heating is continued until charring is observed. At this point, the tube is removed from the heating block and allowed to cool for 5 min. The contents of the tube are treated with 1 mL of 30% hydrogen peroxide, added dropwise, and heating is resumed cautiously. Heating is continued, with additional dropwise additions of 1-mL increments of 30% hydrogen peroxide made as necessary, until digestion is

Table 7.1. Accuracy and Precision Data for Standard Reference Materials (mg/kg ± σ)

	Reference material					
	Cd	Ni	Pb	Cr	Cu	Zn
National Bureau of Standards (NBS) orchard leaves, found	0.15 ± 0.05	1.18 ± 0.08	42 ± 1.7	2.7 ± 0.17	11.2 ± 0.18	25.3 ± 0.5
NBS orchard leaves, certified	0.11 ± 0.1	1.3 ± 0.2	45 ± 3	2.6 ± 0.3	12 ± 1	25 ± 3
NBS bovine liver, found	0.30 ± 0.07	—	0.28 ± 0.04	—	187 ± 2.3	131 ± 1.4
NBS bovine liver, certified	0.27 ± 0.04	—	0.34 ± 0.08	0.088 ± 0.012	193 ± 10	130 ± 13
International Atomic Energy Agency (IAEA) fish flesh, found	0.13 ± 0.04	1.34 ± 1.47	0.40 ± 0.04	2.92 ± 0.13	3.7 ± 0.46	32.4 ± 0.6
IAEA fish flesh, tentative	0.16 ± 0.04	1.2 ± 0.2	0.7 ± 0.2	2.9 ± 0.6	4.6 ± 0.4	36 ± 3

Table 7.2. Statistical Data for Tissues from Contaminated Fish (mg/kg)
(Coefficients of variation, %, are given in parentheses)

Tissue	Cd	Ni	Pb	Cr	Cu	Zn
Rainbow trout, homogenate No. 1	0.10 (12.1)	0.16 (15)	0.92 (4.3)	—	0.6 (13)	10.9 (4.8)
Rainbow trout, homogenate No. 2	0.10 (8.2)	0.15 (13)	0.98 (6.0)	—	0.80 (8.8)	10.9 (3.8)
Rainbow trout, homogenate No. 3	0.11 (9.1)	0.20 (17)	0.94 (8.3)	—	0.53 (6.4)	11.8 (9.3)
Rainbow trout, homogenate No. 4	0.06 (20)	0.21 (8.9)	0.83 (10)	2.2 (7.9)	0.58 (7.1)	12.6 (3.9)
Liver	2.2 (14)	0.92 (11)	8.0 (17)	—	—	—
Kidney	7.1 (9.2)	1.9 (5)	36 (11)	—	—	—
Coho Jack	—	—	—	0.21 (15)	1.06 (5.4)	24.6 (6.7)

complete. The tubes are then cooled and the contents diluted to 50 mL with high-purity water.

For the determination of nickel, lead, and cadmium, a 40-mL aliquot of the digested material is transferred to a 200-mL volumetric flask and diluted with 60 mL of high-purity water. Five mL of 1% w/v aqueous APDC and 5 mL of MIBK are added, and the flask is shaken vigorously for 5 min. The nickel, lead, and cadmium levels are determined from the absorbances obtained by aspiration of the organic phase into the air-acetylene flame. Quantitation is by direct comparison to standards carried through the APDC-MIBK extraction procedure. Specific conditions for the measurements of the absorbances are contained in Sections 5.22.2, 5.16.2, and 5.8.2, respectively.

Zinc, copper, and chromium are determined by aspirating the 10 mL of digested material remaining in the test tube into the air-acetylene flame and measuring the appropriate absorbances. Quantitation is by direct comparison. Sections 5.38.2, 5.12.2, and 5.10.2 contain the specific conditions for the measurement of absorbances from zinc, copper, and chromium, respectively.

7.5. NIOSH Method P&CAM 139 for Arsenic in Urine

NIOSH method P&CAM 139 meets the regulatory compliance monitoring requirements for arsenic exposures in the occupational environment. The organic matrix of the sample is destroyed by acid digestion, and its arsenic content is determined by flameless atomic absorption spectrometry of the gaseous hydride. Six replicate analyses of a urine sample containing 0.4 mg/L on each of six different days showed a coefficient of variation of 11%.

The urine samples are collected in acid-washed, 4-oz polyethylene bottles and preserved by the addition of 0.1 g of EDTA. The specific gravity of the sample should be determined at room temperature at the time of collection.

A 25-mL aliquot of the sample is transferred to a 125-mL beaker and treated with 3 mL of nitric acid and 1 mL of sulfuric acid. The contents of the beaker are heated on a hotplate at 150°C until dense white fumes of SO_3 are evolved. The contents of the beaker are allowed to cool. They are quantitatively transferred to a 25 mL volumetric flask and brought to volume with high-purity water.

A 5-mL aliquot of the prepared sample is transferred from the volumetric flask to the arsine generator flask. Before connecting the generator flask to the hydride generating system, 25 mL of high-purity water and 3 mL of concentrated hydrochloric acid are added. After connecting the flask, a 200-mg sodium borohydride pellet is added via the stopcock. The gaseous hydride is collected in the balloon reservoir and after 1 min, introduced to the argon-hydrogen flame by manipulation of the lower stopcock of the hydride generating system. The absorbance at 193.7 nm is recorded on a strip chart. The arsenic content of the sample is determined by direct comparison. The result is corrected to a urine of specific gravity 1.024 by the multiplicative factor:

0.024/(specific gravity of sample - 1.000)

7.6. NIOSH Method P&CAM 262 for Lead in Blood and Urine

NIOSH method P&CAM 262 describes the regulatory compliance monitoring procedures for lead exposures in the occupational environment. The organic matrix of the sample is destroyed by acid digestion, and the lead is concentrated by APDC-MIBK extraction. The lead level in the sample is determined by direct comparison after aspiration of the organic phase into the air-acetylene flame and measurement of the absorbance at 283.3 nm. The working range of the method is from 5 µg to 250 µg per 100 g of blood and from 10 µg/L to 500 µg/L of urine for sample volumes of 10 mL and 50 mL, respectively. The lead value obtained by analysis of nine replicates of the U.S. NBS SRM No. 2672 freeze-dried urine compared favorably with the certified values, ie:

	Experimental	Certified
Mean (µg/g)	0.99	1.00
Standard deviation	0.064	0.023

Blood samples, 10 mL in volume, are collected in lead-free, heparinized vacutainers* using sterile stainless steel

---

*Becton-Dickinson and Co., Rutherford, N.J. (#4610 Vacutainer).

needles. The samples may be stored for up to 2 weeks under refrigeration.

A 50-mL urine sample is collected in lead-free, borosilicate or polyethylene bottle. The specific gravity of the sample is recorded at room temperature at the time of collection. The urine sample is preserved with thymol (5 mg per 100 mL of urine), and it may be stored for up to 1 week under refrigeration.

The sample, 10 g of blood or 50 mL of urine, is measured into a 125-mL beaker, treated with 6 mL of concentrated nitric acid, covered with a watch glass, and heated overnight at 120° to 140°C on a hotplate. The samples are then evaporated to near dryness. If a nearly white residue is not obtained, more nitric acid is added and the evaporation step is repeated. The residue is treated with 1 mL of nitric acid and 5 mL of high-purity water and heated. When the residue has dissolved, the solution is transferred to a 50-mL volumetric flask. The contents of the flask are treated with 2 drops of phenol red indicator (0.1% w/v) and 5 mL of ammonium citrate buffer (400 g citric acid cautiously dissolved in 500 mL of ammonium hydroxide, followed by adjustment to pH 8.5). Ammonium hydroxide is added until the indicator turns red. The contents of the flask are then treated with 1 mL of 1% w/v potassium cyanide solution, 1 mL of 2% w/v APDC solution, and 4 mL of water-saturated MIBK. The flask is shaken for 30 sec, the phases are allowed to separate, and the MIBK layer is floated up into the neck of the flask by the addition of high-purity water.

The organic phases containing the samples and standards are aspirated into the air-acetylene flame, and their absorbances at 283.3 nm are recorded. The lead content of the samples is determined by direct comparison.

The urinary lead concentrations are converted to specific gravity 1.024 by the multiplicative factor:

0.024/(specific gravity of sample - 1.000)

7.7. NIOSH Method P&CAM 167 for Mercury in Blood

NIOSH method P&CAM 167 is applicable to the determination of blood mercury levels for regulatory compliance monitoring

of mercury exposure in the occupational environment. Blood
samples are prepared by low-temperature digestion with sulfuric acid and potassium permanganate, and the mercury content
of the blood is determined by the cold vapor technique.

Blood samples are collected in heparinized vacutainers
using sterile stainless steel needles. The samples may be
stored for up to 2 weeks under refrigeration.

A 2-mL aliquot of the blood sample is pipetted into a stoppered 125-mL Erlenmeyer flask and diluted with 5 mL of high-
purity water. The contents of the flask are treated with 5 mL
of sulfuric acid, and the flask is stoppered and incubated in
a shaker water bath maintained at 54°C for 3 h. The flask is
then transferred to an ice bath, and its contents are treated
with 17 mL of 6% (saturated) potassium permanganate solution.
The flask is stoppered and allowed to stand overnight at room
temperature. After this time, the contents of the flask are
treated with 2 mL of 20% w/v hydroxylamine hydrochloride solution, transferred to a 100-mL volumetric flask, and brought to
volume with high-purity water.

The mercury content of the prepared sample is measured by
the method of standard additions. The absorbances of a 20-mL
aliquot of the prepared sample treated with 1 mL of high-
purity water and of three 20-mL aliquots spiked with 1 mL of
either a 0.05-µg/mL, a 0.25-µg/mL, or a 0.50-µg/mL standard
mercury solution are measured at 253.7 nm. The sample or
spiked sample, contained in the reaction vessel of the cold
vapor generating apparatus, is treated with 1 mL of 20% stannous chloride solution (20 g dissolved in 100 mL of 6 N hydrochloric acid), and the vessel is immediately connected to the
generating train for the measurement of absorbance.

7.8. NIOSH Method P&CAM 165 for Mercury in Urine

The method described in this section is applicable to the
determination of urinary mercury levels for purposes of compliance monitoring of exposures in the occupational environment under the Occupational Safety and Health Act. Urine
samples are prepared by acid digestion, and their mercury content is measured by the cold vapor technique.

Samples of at least 25 mL are collected in acid-cleaned
borosilicate glass bottles and preserved with 0.1 g of

potassium persulfate. Samples so preserved are stable for
2 weeks at room temperature.

The method of standard additions is employed to determine
the mercury level of the urine sample. One-milliliter aliquots of the sample are treated with either 1 mL of water or
1 mL of mercury standard in concentrations of 0.05 µg/mL,
0.1 µg/mL, or 0.15 µg/mL. The sample and spiked samples are
treated with 5 mL of nitric acid and allowed to stand at room
temperature for 3 min. They are then transferred to the reaction vessel of the cold vapor generating apparatus, diluted
to 100 mL with high-purity water, treated with 1 mL of 20% w/v
stannous chloride solution, and immediately connected to the
generating train. The absorbances at 253.7 nm are recorded.

8. QUALITY ASSURANCE

8.1. Role of Quality Assurance

A tremendous resource of environmental data is being generated annually by governmental, commercial, and industrial
laboratories. These data are utilized for the assessment of
the toxic and asthetic qualities of drinking water, the determination of the acceptability of water and wastewater treatment, and also the development of water quality planning and
management strategies. The importance of the decisions that
are made based upon these data require that they be valid,
defensible data of known quality.

Regulatory agencies have been placing more and more empha-
sis on quality assurance programs as a mechanism for identifying and documenting data quality. Quality assurance should
not be confused with quality control. For purposes of this
writing, quality control is defined as those activities performed on a day-to-day basis to insure that the data being
generated are valid. Quality assurance is defined as an independent assessment of the monitoring program to insure that
the overall process (ie, sample collection, sample analysis,
quality control) that has been developed to generate valid
data is functioning properly. Many state agencies operate
laboratory certification programs as a type of quality assurance program that assesses and documents the quality of self-
compliance monitoring data. In instances where the quality of
these data is poor, strict regulations are often adopted to
establish minimum standards and criteria for laboratories to
follow when generating these data.

Another regulatory approach to identifying data quality is
requiring monitoring projects to have a written quality assurance project plan. This document identifies the quality assurance objectives of the project, sample collection and handling procedures, analytical methodologies to be used, quality
control practices to be implemented, and the data handling and
validation procedures to be followed.

In any case, regulatory agencies have begun addressing this
problem of identifying data quality through the implementation

of quality assurance programs. These programs operate as an
independent assessment of the monitoring projects and cover
all aspects from sample collection to reporting of the analytical data.

When developing any monitoring project, it is important to
identify and document the:

1. Monitoring project objectives

2. Project personnel responsibilities and organization

3. Quality assurance objectives

4. Sampling procedures to be used

5. Analytical methodologies to be used

6. Instrumental calibration procedures

7. Quality control procedures to be followed

8. Procedure for data reduction and reporting

By performing this task, the process for monitoring the quality of data being generated and the responsibility of insuring
that the objectives of the quality assurance program are met
are clearly defined and documented. Then throughout the monitoring project, the individual assigned the responsibility
of monitoring the quality of the data, the quality assurance
coordinator, should review all aspects of the monitoring process. As sampling or analytical problems arise, the quality
assurance coordinator should qualify or reject the data that
the problem has impacted, review the problem, recommend corrective action, and monitor the problem closely to determine
whether the corrective action has been effective in alleviat-
ing the problem. This process identifies invalid data and
assures the monitoring project that the data generated are of
the quality required for the monitoring objectives.

This chapter reviews the major points of a quality assurance program and discuss the important considerations that
should be made when attempting to generate valid, defensible
data.

8.2. Monitoring Project Description
and Personnel Objectives

When generating data for any monitoring project, it is essential that all personnel working on the project have an understanding of the purpose of the project, their function in relation to the overall monitoring process, and the responsibility, as well as authority placed with all individuals participating on the project. This may be accomplished by writing a brief project description and preparing an organizational chart of the overall project. In addition, the responsibility and authority of each individual must be clearly defined. If and when a problem does arise during the project, this will enable personnel to immediately notify the appropriate authority so that corrective action can be initiated and that any data adversely effected by the problem can be qualified as "questionable results" or rejected.

8.3. Environmental Sampling

8.3.1. Sample Collection

An integral part of any monitoring project, be it biological, waste water, potable water, ground water or surface water, is planning. Depending upon the type of environmental system being monitored (ie, stream, ambient air), the presurvey plan should first address:

1. What is the objective of the monitoring project?

2. What type of sample is to be collected?

3. What method of sample collection is to be employed?

The objective of the monitoring project will dictate the type of sample to be collected (ie, water, soil, air). However, as discussed in Chapter 2, there are many methods of sample collection that may be used. The major factor that must be considered when selecting the method(s) of sample

collection is that the objective of sampling is to collect a
representative portion of the total environmental system being
monitored. Other factors that should be considered are man-
power resources available and sampling site location. From
this information, the project planner should then decide the
frequency of sampling needed for the assessment, the specific
sampling sites to be monitored, and the method of sample collection to be used (ie, grab or composite).

8.3.2. Sample Labeling

All samples submitted to a laboratory for analysis should
be properly labeled and accompanied by an analysis request
form. The information put on labels and forms, should be
printed legibly. Cross outs or erasures should not appear on
the labels or forms and if mistakes are made, the label or
form should be discarded and a new form filled out.

Sample labels should be attached (by adhesives, elastic
straps, etc) to the sample bottle and should indicate:

1. Sample bottle number

2. Type of sample (stream, discharge, etc)

3. Preservative used

The sample analysis request form should indicate:

1. Sampling site location

2. Sample collector

3. Date and time of sample collection

4. The corresponding sample bottle number

The information should be filled out on the label and form
at the time of sample collection and should accompany the samples to the laboratory. When submitting samples to the laboratory, the sample collector should retain a copy of the
analysis request form as a record of sample submittal.

8.3.3. Chain of Custody

All monitoring programs being conducted for compliance with

state or federal regulations or programs involving data that
may be used in a court of law should document and implement a
chain of possession and custody of any sample collected. This
procedure is to assure and document that at all times the sample was in possession of a specific individual (physical possession or a secured space) and could not have been damaged
or altered in any way.

A chain of custody form, if used, should list at a minimum
the following information:

1. Sample bottle number

2. Description of sample

3. Specific location of sample collection

4. Signature of sample collector

5. Date and time of sample collection

6. Sample preservation steps taken

Once the sample is submitted to the laboratory, the following information should be added to the chain of custody form:

1. Date and time of custody transfer to the laboratory (if
   the sample was collected by a person other than laboratory personnel)

2. Signature of person accepting custody (if the sample was
   collected by a person other than laboratory personnel)

3. Date and time of initiation of analysis

4. Signature of person performing the analysis

5. Name of laboratory performing the analysis

Upon completion of the analysis, this document should be
returned with the reported results to the sample collector.

8.3.4. Sample Handling and Preservation

As discussed in Chapter 2, careful consideration should be
given to the type of sample container to be used for sample
collection and transport to the laboratory. Glass or hard
plastic (polyethylene with a polypropylene cap with no liner)[9]

is recommended for trace metal samples.

Preservation techniques are designed to retard the chemical and biological changes that may occur in a sample once it has been removed from its source (ie, in water samples, metal cations may precipitate as hydroxides or form complexes with other constituents).

As discussed in Chapter 2, sample preservation in some sample matrices, such as sludge or sediment samples, may not be easily accomplished. In those instances, the sample should be refrigerated and immediately transported to the laboratory and analyzed as soon as possible.

When collecting potable water, waste water, or other water-type samples, samples should be preserved with nitric acid to a pH < 2 immediately following sample collection. This may be accomplished by adding the acid (normally 3 mL of 1:1 nitric acid per liter of sample) to the sample bottle in the laboratory prior to sample collection or by adding the acid to the sample immediately following sample collection. The following is a recommended procedure for the latter case:

1. Add approximately 2 mL of 1:1 nitric acid to sample immediately following sample collection.

2. Replace stopper or cap on sample bottle and mix sample thoroughly by inverting bottle several times.

3. Remove sample bottle stopper and place a drop of sample from the stopper onto pH test paper.

4. Rinse the portion of the stopper exposed to the pH test paper with distilled or deionized water.

5. If pH is not less than 2, repeat steps 1 through 5.

Note: Care should be taken to follow the above steps. Adding too much preservative, "over preservation," will dilute the sample and yield inaccurate results.

After the samples have been properly collected and preserved, they should be transported to the laboratory and analyzed as soon as possible. Federal regulations require that properly preserved drinking water samples be analyzed within 6 months from the time of sample collection for trace metals (excluding mercury), 38 days for samples collected in glass containers for mercury analysis, and 13 days for samples collected in plastic containers for mercury analysis.[9] The U.S. Environmental Protection Agency (EPA) also recommends that the above holding times be applied to wastewater analysis, except the holding for mercury in a plastic container is reduced to 13 days. The New Jersey State Department of Environmental Protection requires that the above holding times be applied to all samples being analyzed for compliance with the New Jersey Safe Drinking Water Act and the New Jersey Water Pollution Control Act.[27]

8.3.5. Field Logbooks

Each sample collector should have a bound daily logbook. This book should accompany the sampler in the field and at each sampling site, the sample collector should record:

1. Date and time of sampling

2. Specific sampling site

3. Sample bottle number

4. Field measurements (if conducted)

5. Weather conditions (if applicable to the survey)

6. Special comments (pertaining to the sample or sampling site, etc)

All field logbooks should be neat, entries should be in ink and legible. The sample collector should use the field logbook as a record of what samples were collected and submitted to a laboratory, keeping in mind that this logbook may be used in a court of law at some future date.

8.3.6. Quality Control

Quality control should be included as part of the field aspects of monitoring as well as the laboratory aspect. Field quality control should address sampling equipment, field measurements, and sample collection procedures. A general rule of thumb is that 5% of the field monitoring efforts should be directed toward quality control.

Procedures for calibration and maintenance of field

instruments and sampling devices should be developed, implemented, and documented. These procedures would apply to
stream flow monitoring devices, automatic samplers, and other
automated equipment. This documentation should address:

1. Description of the cleaning procedure to be used on the
   equipment

2. Description of the calibration procedure to be used on
   the equipment

3. Frequency of cleaning and calibration of the equipment

Also, if other field measurements are to be performed in
conjunction with the survey (ie, pH, dissolved oxygen), the
procedures for calibrating and maintaining that instrumentation should be documented.

The following samples should be collected and analyzed for
the same contaminants being monitored in the survey as part of
the quality control check on the monitoring activities:

1. Sample preservative blanks. Sample preservatives, if
   used, may become contaminated during field usage.
   Therefore, the same quantity of acid normally added to
   samples should be added to a sample bottle filled with
   laboratory pure water and this sample should be submitted to the laboratory as a check on preservative
   contamination.

2. Automatic sampler blanks. If automatic samplers are
   used to collect the samples, a quantity of laboratory
   pure water should be passed through the automatic sampler just after the sampler has been cleaned. This sample should also be submitted to the laboratory for analysis as a check on the cleaning procedure efficiency.

3. Duplicate samples. Duplicate samples should be collected at selected stations using two sets of sampling equipment at the same site or duplicate grab samples. These
   samples should be submitted to the laboratory for anal-
   sis as a check on the sampling equipment and technique
   for precision.

4. Split samples. Two representative subsamples are

removed from one collected sample and submitted to the laboratory for analysis. These data may be used as a check on the precision of the analytical precedure used in the laboratory.

8.4. Laboratory Analysis

The responsibility of the analytical laboratory is to provide a qualitative and quantitative analysis of the sample being submitted. In order to accomplish this, it is imperative that specific steps be taken within the laboratory to enable the analyst to determine whether that particular analytical process is "in control." The following sections will address these steps.

8.4.1. Sample Receipt

When samples are submitted to a laboratory, laboratory personnel should check the integrity of the samples. That is, the samples should be checked for:

1. Proper identification (ie, site of sample collection and collector's name)

2. Date and time of sample collection

3. Whether the sample has been properly preserved

4. Sufficient sample volume for the analysis requested

5. Proper sample container.

It is important to check these items in order to avoid problems which may arise later during the analytical process. When checking the date and time of sample collection, the analyst should check to determine whether any regulated sample holding times (refer to Chapter 2) will be violated. When checking the acid or base preservation of a sample, the sample receiver should analyze a portion of that sample with a pH meter and document whether the sample was properly preserved. The sample receiver should also check the sample volume to determine whether there is enough sample present not only to conduct the analysis but also to meet the quality control required.

Upon checking the integrity of the sample, the sample receiver should document the acceptability of the sample. Samples improperly preserved, exceeding established holding times,
or collected in improper sample containers should be rejected
by the sample receiver or the data that are generated should be
qualified as "questionable results." The sample receiver
should keep in mind that once the sample is received, the laboratory assumes full responsibility for that sample.

8.4.2. Sample Analysis

The selection of the analytical method to be used for a
particular analysis should be based upon the monitoring objectives. If the sample has been collected as part of a state
or federal compliance monitoring requirement, the method of
sample analysis may have been previously designated by that
agency (refer to Chapters 4, 5, and 6).

The method of choice should be well documented by the laboratory. Some state and federal regulatory agencies require
laboratories to develop and maintain laboratory methods manuals. These manuals describe, in detail, the analytical methods for all of the analyses performed by the laboratory.
These manuals provide an historical record of the exact methods employed by the laboratory and indicate the specific op-
tions taken (if provided in the method) or deviations made (if
legally allowed) when conducting that particular analysis.

8.4.3. Quality Control

As stated earlier, quality control is defined as those activities performed on a day to day basis to insure that the
data being generated are valid. This should take up, at a
minimum, 10% of the laboratory's efforts. This includes not
only the conducting of spiked and replicate analyses, but also
performing checks and tests on the instrumentation and reagents used in the course of the analytical process. The following are quality control checks specific to atomic absorption
spectrometry that should be performed and documented:

8.4.3.1. Instrument Quality Control

1. Atomic absorption spectrophotometer. After following

the manufacturer's instructions for adjusting and calibrating the atomic absorption spectrophotometer, the instrument's maximum obtainable sensitivity should be
checked for that particular element by analyzing a specific standard and recording the standard's concentration,
absorption reading, date of check, and analyst in a bound
notebook. This process should be repeated each time the
instrument is set up for a particular analysis. Reduction in the instrument's sensitivity may be indicative
of an instrumental malfunction or aging of the light
source and, therefore, require prompt corrective action.

2. Conductivity meter. The conductivity meters used for
   checking the purity of the laboratory water (as discussed
   below) should be checked annually. Meters equipped with
   conductivity cells having platinum electrodes should be
   checked over the range of interest using at least five
   concentrations of a standard potassium chloride solution.
   Meters not equipped with cells having platinum electrodes
   should be checked against a conductivity meter equipped
   with platinum electrodes. When this check is performed
   the raw data, cell constant, comparison results, date of
   check, analyst, and correction factor (if needed) should
   be recorded in a bound notebook.

3. Analytical balance. If an analytical balance is used in
   the preparation of calibration standards, the balance
   should be checked and adjusted annually by a service
   person employed by the laboratory or balance consultant.
   The accuracy of the balance should be checked once a
   month using at least two class "S" weights. The weights
   used in the once a month check, weight detected to the
   nearest 0.1 mg, dates on which the checks were performed,
   analyst, and other pertinent information should be recorded in a bound notebook.

8.4.3.2. Reagent Quality Control

1. Laboratory pure water. To avoid the introduction of
   contaminants into the analyses by way of the laboratory
   pure water that may be used (refer to Section 3.3.1),
   the purity of the water should be checked daily using a
   conductivity meter. The conductivity reading, date of

check, and analyst should be documented in a bound notebook. Preferably, laboratory pure water should have a
conductivity value of less than 1 μmho/cm.

2. Reagents and chemicals. Spectroquality chemicals should
   be used for trace metal analysis, although sometimes reagent grade quality may be satisfactory. A reagent blank,
   as discussed below, should be analyzed to determine whether a particular reagent contains a contaminant or chemical that may interfere with a particular analysis. The
   following practices should be followed:

   a. All chemicals, solutions, and standards should be
      dated upon receipt.

   b. All solutions should be properly labeled with the
      identification of the compound, concentration, date
      of preparation, and analyst who prepared the solution.

   c. Stock and working solutions should be checked regu-
      larly for signs of decomposition which may be indi-
      cated by discoloration, formation of precipitates,
      and concentration change due to evaporation.

8.4.3.3. Analytical Quality Control

1. Reagent blanks. Each individual reagent used in the
   analytical procedure should be tested to determine whether it causes any interference with the analysis. The
   conditions for handling and analyzing the blank should
   be identical to that used in the analysis. The reagent
   blank should be analyzed with each new reagent and the
   data generated from the reagent blank should be documented.

2. Method blank. The method blank is analyzed to determine
   whether the cumulative effect of the reagents causes
   interference with the analysis. The method blank
   should consist of only laboratory pure water and the
   reagents used in the analysis. The method blank is
   handled and analyzed in the same manner as standards
   and samples. A method blank should be analyzed each
   time an analysis is conducted.

3. Replicate analyses. A minimum of 5% of all samples

being analyzed should be tested in replicate. Replicate
samples are prepared by dividing a homogeneous sample
into separate parts so that each part is also homogeneous
and representative of the original sample. The data obtained from the replicate analyses should be used to
document the precision of that particular method.

4. Spiked analyses. A minimum of 5% of all samples being
analyzed should be analyzed as spiked samples. Spiked
analyses are performed by splitting a sample into replicates and to one of the replicates, a known amount of
the contaminate being tested for is added. The amount
of the contaminate being added should be approximately
the same amount present in the unspiked sample. Both
samples should then be analyzed and the percentage recovery of the spike may be expressed as:

$$R = 100 \ (F-I)/A$$

where F is the analytical result of the spiked sample,
I is the result before spiking of the sample, and A is
the amount of contaminant added to the sample. This
information should be documented.

5. Standard deviation. The standard deviation (s) should
be calculated and documented for each parameter being
analyzed. This may be accomplished by analyzing a specific concentration of a standard and documenting the
analytical result. After 20 determinations have been
made, the standard deviation may be calculated using the
following equation:

$$S = \sqrt{\frac{\Sigma (x-\bar{x})^2}{n-1}}$$

where Σ is the summation of values, x is the observed
value, $\bar{x}$ is the mean or average value, and n is the number of observations. Example: Over a period of 2 months,
a 5-ppm copper standard was analyzed 20 times and the
standard deviation calculated (Table 8.1):

Table 8.1. Standard Deviation Calculation

Observation	Detected value	$(x-\bar{x})$	$(x-x)^2$
1	4.89	-0.10	0.01
2	5.01	0.02	0.0004
3	4.98	-0.01	0.01
4	4.95	-0.04	0.0016
5	5.03	0.04	0.0016
6	5.03	0.04	0.0016
7	5.14	0.15	0.0225
8	5.03	0.04	0.0016
9	4.95	-0.04	0.0016
10	5.01	0.02	0.0004
11	5.02	0.03	0.0009
12	4.98	-0.01	0.0001
13	4.97	-0.02	0.0004
14	5.04	0.05	0.0025
15	5.05	0.06	0.0036
16	5.01	0.02	0.0004
17	4.93	-0.06	0.0036
18	4.90	-0.09	0.0081
19	4.98	-0.01	0.0001
20	4.96	-0.03	0.0009

$\bar{x} = 4.99$ $\qquad\qquad\qquad$ $\Sigma(x-\bar{x})^2 = 0.0719$

$$S = \sqrt{\frac{\Sigma(x-\bar{x})^2}{n-1}} = 0.06$$

Once calculated, the standard deviation can be expressed using a quality control chart to determine whether an analyst is in control of the analyses. The control chart in Figure 8.1 is a plot of the standard's mean value $(\bar{x})$, observed value, upper control limit (UCL), and the lower control limit (LCL). Typically, the UCL and LCL are based upon two standard deviations from the standard's mean value. Once the control chart is generated for a particular analysis, the standard used

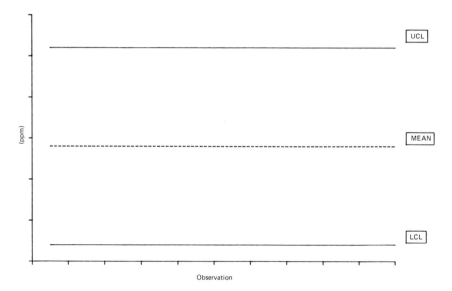

Figure 8.1. Basics of a control chart.

to calculate the standard deviation is analyzed with each future analytical run. The value determined for that standard's
analysis is then plotted on the control chart. If the value
determined exceeds either the UCL or the LCL, the analysis is
termed "out of control" and the analyst should then begin
checking the procedure for the error. Once the problem is located and corrected (ie, instrumental failure or replacing of
reagents), the analyst should rerun the standard to verify the
problem has been corrected. Figure 8.2 is an example of a
control chart using the data listed in Table 8.1. Quality control charts can be a simple graphical means for analysts to
document and review quality control data.

8.4.4. Data Handling and Documentation

Once the analysis of the sample is completed, a record
should be made in the analyst's notebook of:

1. Date and time of analysis

2. Analytical method employed

3. Raw data of standards

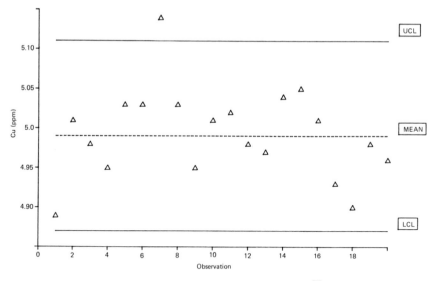

Figure 8.2. Example of a control chart. $\overline{x} = 4.99$; s = 0.06; UCL = 4.99 + 2s = 4.99 + 0.12 = 5.11; LCL = 4.99 - 2s = 4.99 - 0.12 = 4.87.

4. Quality control raw data

5. Sample number, raw data, and final result

6. Calculations used to determine final result

7. Analyst's signature

A review process should also be incorporated into the laboratory procedure. That is, a second analyst should review the first analyst's quality control data, raw data, and calculations to verify the reported results. Upon completion, the second analyst should also initial the notebook. The data may now be reported to the sampling agency and the reporting process should be checked for transcribing errors.

All laboratory data (ie, raw analytical data, quality control data, sample reports) should be well documented and these records should be maintained as a permanent record for many years. If at any time the validity or nature of the results is questioned, these records are needed to verify the results.

8.4.5. Chain of Custody

As discussed in Section 8.3.3, in some instances the data which are generated may be used in a court of law. If so, the laboratory should establish a set of protocols designed to insure and document the integrity and custody of the samples.

The laboratory's chain of custody procedures should begin with the receipt of the samples and continue through the analytical process. The chain of custody form used (should be the same form stated in Section 8.3.3) should indicate at a minimum the following:

1. Date and time the laboratory received custody of the sample

2. Person accepting custody of the sample

3. Whether the sample was received preserved or unpreserved

4. Date and time of analysis of the sample

5. Person or persons who performed the analysis

6. Type of analysis performed and the analytical method employed

Once completed, the chain of custody form should be attached to the analytical report and forwarded to the sampling agency.

8.5. Quality Assurance Documents

This chapter has stressed two main points which should be addressed whenever a monitoring program is being designed. These are (1) defining each step of the monitoring process and (2) documenting it prior to and during the process. The amount of work involved in performing these two tasks may be minimized through the development of standard operating procedures (SOP) manuals. A field SOP and a laboratory SOP should be developed and maintained through the monitoring program. These manuals document the routine procedures practiced and may serve as a guide or instruction manual to all personnel.

Once the SOPs are established, the routine quality control record keeping practices should be developed (ie, notebooks

for each analyst, instrument). If a routine recording procedure is developed, the amount of time devoted to maintaining
the records is then kept to a minimum and the only major cost
associated with quality control is the quality control analyses, review of the quality control data, and implementation of
the data when a problem arises. This cost is usually 10% to
15% of the overall monitoring program costs. This cost is considered reasonable when compared to the consequences that may
occur based on decisions made on invalid data.

REFERENCES

1. Baird Corp. "Atomic Absorption Spectrometer." Baird Corporation, Bedford, MA, 1979.

2. Instrumentation Laboratory, Inc. "This Is IL." Instrumentation Laboratory, Inc., Wilmington, MA, 1979.

3. Perkin-Elmer Corp. "The Guide to Techniques and Applications of Atomic Spectroscopy." Perkin-Elmer Corp., Norwalk, CT, 1980.

4. Welz, B. "Atomic Absorption Spectrometry." Verlag Chemie, Weinheim, 1976, pp. 133-181.

5. Van Loon, J. C. "Analytical Atomic Absorption Spectroscopy: Selected Methods." Academic Press, New York, 1980, pp. 77-311.

6. Perkin-Elmer Corp. "Analytical Methods for Atomic Absorption Spectrometry." Perkin-Elmer Corp., Norwalk, CT, 1973.

7. Perkin-Elmer Corp. "Analytical Methods for Atomic Absorption Spectrometry." Perkin-Elmer Corp., Norwalk, CT, 1976.

8. U.S. EPA. "Manual of Methods for Chemical Analysis of Water and Wastes." U.S. Environmental Protection Agency, Washington, DC, 1974.

9. U.S. EPA. "Manual of Methods for Chemical Analysis of Water and Wastes." U.S. Environmental Protection Agency, Washington, DC, 1979.

10. NIOSH. "Manual of Analytical Methods." National Institute of Occupational Health and Safety, Washington, DC, 1979.

11. Code of Federal Regulations, Title 40, Part 50, Appendix G. "Reference Method for the Determination of Lead in Suspended Particulate Matter Collected from Ambient Air." U.S. Government Printing Office, Washington, DC, 1980, pp. 654-659.

12. U.S. EPA. "Handbook for Monitoring Industrial Wastewater." U.S. Environmental Protection Agency, Washington, DC, 1973, pp. VI.1-VI.18.

13. Canadian DOE. "Analytical Methods Manual." Canadian Department of the Environment, Ottawa, 1979.

14. U.S. EPA. "Test Methods for Evaluating Solid Waste, Physical/Chemical Methods." 2nd ed. U.S. Environmental Protection Agency, Washington, DC, May, 1982.

15. ASTM. "1980 Annual Book of ASTM Standards: Part 31, Water." American Society for Testing and Materials, Philadelphia, PA, 1980.

16. ASTM. "1975 Annual Book of ASTM Standards: Part 31, Water." American Society for Testing and Materials, Philadelphia, PA, 1975.

17. Rand, M.C., Taras, M.J., and Greenberg, A.E. "Standard Methods for the Examination of Water and Wastewater," 14th ed. American Public Health Association, Washington, DC, 1975.

18. Katz, S.A. "Inorganic Composition of the Near Shore Waters of the Southern New Jersey Coast." Annual Meeting of the New Jersey Academy of Science, Madison, NJ, April 1978.

19. Kolthoff, I.M., Sandell, E.B., Meehan, E.J., and Bruckenstein, S. "Quantitative Chemical Analysis." Macmillan Co., Toronto, 1969, pp. 516-518.

20. Rieman, W., Neuss, J.D., and Naiman, B. "Quantitative Analysis." McGraw-Hill, New York, 1951, pp. 13-14.

21. Valentine, J.L., Kang, H.K., and Spivey, G. Environ. Res. 1979, 20, 24-32.

22. Knause, D., and Katz, S.A. "Lead Levels in Roadside Vegetation." Annual Meeting of the New Jersey Academy of Science, Union, NJ, April 1977.

23. Clanet, P., DeAntonio, S.M., Katz, S.A., and Scheiner, D.M. "Effects of Some Cosmetics on the Levels of Selected Trace Metals in Human Scalp Hair." 64th Conf. Canadian Inst. Chem., Halifax, June 1981.

24. Chuang, L.S., Kwong, L.S., and Yeh, S.J. J. Radioanal. Chem. 1979, 49, 103-113.

25. Mamuro, T., Matsuda, Y., and Mizohata, A. Ann. Rept. Radiation Cent. Osaka Prefecture 1971, 12, 10-12.

26. Katz, M., Ed. "Methods of Air Sampling and Analysis." American Public Health Association, Washington, DC, 1981.

27. NJ DEP. "Field Procedures Manual for Water Data Acquisition." New Jersey Department of Environmental Protection, Trenton, NJ, 1980, pp. 3-15.

28. Wagstaff, D.J., Brown, J.F., and McDowell, J.R. FDA By-Lines 1979, 9, 271-286.

29. U.S. FDA. Laboratory Information Bulletin No. 2264, U.S. Department of Health, Education, and Welfare, Washington, DC, 1979.

30 Moody, J.R., and Lindstrom, R.M. Anal. Chem. 1977, 49, 2264-2267.

31. Laxen, D.P., and Harrison, R.M. Anal. Chem. 1981, 53, 345-350.

32. U.S. EPA. "Handbook for Analytical Quality Control in Water and Wastewater Laboratories." U.S. Environmental Protection Agency, Washington, DC, 1972.

33. Zief, M., and Mitchell, J.W. "Contamination Control in Trace Element Analysis." John Wiley and Sons, New York, 1976, p. 75.

34. U.S. EPA. "Investigation of Matrix Interferences for AAS Metal Analysis of Sediments." U.S. Environmental Protection Agency, Washington, DC, 1978.

35. Nackowski, S.B., Putnam, R.D., Robbins, D.A., Varner, M.O., White, L.D., and Nelson, K.W. Amer. Indust. Hyg. Assoc. J. 1977, 30, 503-508.

36. Lecomte, R., Parades, P., Monaro, S., Barrette, M., Lamoureux, G., and Menard, H.A. Internat. J. Nucl. Med. Biol. 1979, 6, 207-211.

37. Handy, R.W. Clin. Chem. 1979, 25, 197.

38. Gorsky, J. E., and Dietz, A.A. Clin. Chem. 1978, 24, 169.

39. Lee, R. E., and Duffield, F.V. Sources of Environmental-
    ly Important Metals in the Atmosphere. In "Ultratrace
    Metal Analysis in Biological Sciences and Environment."
    T.H. Risby, ed. American Chemical Society, Washington,
    DC, 1979, p. 148.

40. Pierce Reviews, 1981, 81, 11 (Pierce Chem. Co., Rockfield,
    IL).

41. Bowen, H.M.J. "Trace Elements in Biochemistry." Aca-
    demic Press, New York, 1976, p. 63.

42. Federal Register. "Hazardous Waste Monitoring System."
    General Federal Register, Vol. 45, No. 98, pp. 33075-
    33127, May 19, 1980.

43. Kratochvil, B., and Taylor, J.K. Anal. Chem. 1981, 53,
    924A-928A.

44. Bendix Corp. "Sampling for Toxic Substances." Bendix
    Corporation, Largo, FL, 1981.

45. Wilkerson, C.L., Wehner, A.P., and Rancitelli, L.A.
    Fd. Cosmet. Toxicol. 1977, 15, 589-593.

46. Versieck, J.M.J., and Speeck, A.B.H. "Contamination
    Induced by Collection of Liver Biopsies and Human Blood."
    Proceedings of the Symposium on Nuclear Activation Tech-
    niques in the Life Sciences, Bled, 1972, IAEA, Vienna,
    1972.

47. Masironi, R., and Parr, R.M. "Collection and Trace Ele-
    ment Analysis of Post Mortum Human Samples." The IAEA/
    WHO Research Programme on Trace Elements in Cardiovascular
    Diseases, International Workshop on Biological Specimen
    Collection, Luxembourg, April 1977.

48. Thompson, R.J. Collection and Analysis of Airborne
    Metallic Elements. In "Ultratrace Metal Analysis in
    Biological Sciences and Environment," T.H. Risby, ed.,
    American Chemical Society, Washington, DC, 1979, p. 59.

49. Subramanian, K.S., Chakrabarti, C.L., Suelras, J.E., and
    Maines, I.S. Anal. Chem. 1978, 50, 444-448.

50. Truitt, R.E., and Weber, J.H. Anal. Chem. 1979, 51,
    2057-2059.

51. Hoyle, W.C., and Atkinson, A.   Appl. Spectrosc. 1979, 33, 37-40.

52. Das, H.A., Faahof, A., Gouman, J.M., and Ooms, P.C.A. J. Radio, Anal. Chem. 1980, 59, 55-62.

53. Zdrojewski, A., Quickert, N., Dubois, L., and Monkman, J.L. Internat. J. Environ. Anal. Chem. 1973, 2, 63.

54. Quickert, N., Zdrojewski, A., and Dubois, L.   Internat. J. Environ. Anal. Chem. 1973, 2, 331.

55. Janssens, M., and Dams, R.   Anal. Chim. Acta 1973, 65, 41.

56. Severs, R.K., and Chambers, L.A.   Arch. Environ. Health 1972, 25, 139.

57. Gallorini, M., Genova, M., Meloni, S., and Maxia, V. Inquinamento 1973, 15, 13.

58. Szivos, K., Polos, L., Bezur, L., and Pungor, E.   2nd Internat. Symp. Anal. Chem., Ljubiana, June 1972.

59. Poldoski, J., Leonard, E.N., Fiandt, J.T., Anderson, L.E., Olson, G.F., and Glass, G.E.   Internat. Assoc. Great Lakes Res. 1978, 4, 206.

60. U.S. EPA.   "Sampling and Analysis Procedures for Screening of Industrial Effluents for Priority Pollutants." U.S. Environmental Protection Agency, Washington, DC, April 1977.

61. U.S. EPA.   "Interim Method for the Analysis of Elemental Priority Pollutants in Sludge." U.S. Environmental Protection Agency, Cincinnati, December 1978.

62. Khalily, H.   "Analysis of Severn Estuary Sediments." Ph.D. Thesis, University of Bristol, 1975.

63. Wisseman, R.W., and Cook, S.F.   Bull. Environ. Contamin. Toxicol. 1977, 18, 77-82.

64. Theis, T.L., Westrick, J.D., Hsu, C.L., and Marley, J.J. J. Water Pollut. Control Fed., 1978 (November), 2454-2469.

65. Agemian, H., and Chau, A.S.Y.   Anal. Chim. Acta 1975, 80, 61-66.

66. Ammons, B. Proc. Montana Acad. Sci. 1980, 39, 117-120.

67. Knechtel, J.R., and Fraser, J. Anal. Lett. 1974, 7, 497-504.

68. Carrondo, M.J.T., Lester, J.N., and Perry, R. Talanta 1979, 23, 929-933.

69. Tabatabai, N.A., and Frankenberger, W.T. "Chemical Composition of Sewage Sludges in Iowa." Res. Bull. 586, Agriculture and Home Economics Experiment Station, Iowa State University, Ames, IO 1979.

70. Martin, T.D., Kopp, J.F., and Ediger, R.D. Atomic Abs. Newslett. 1975, 14, 109-116.

71. Hafez, A.A.R., Brownell, J.R., and Stout, P.R. Communic. Soil Sci. Plant Anal. 1973, 4, 333.

72. Khan, H.L., Fernandez, F.J., and Slavin, S. Atomic Abs. Newslett. 1972, 11, 42.

73. Rains, T.C., and Menis, O. J. Assoc. Off. Anal. Chem. 1972, 55, 1339.

74. Price, W.J. "Analytical Atomic Absorption Spectrometry." Heyden and Son, Ltd., London, 1974, p. 160.

75. Mutsch, F., Horak, O., and Kinzel, H. Z. Pflanzenphysiol. 1977, 94, 1-10.

76. George, M.D., and Kureisky, T.W. Indian J. Mar. Sci. 1979, 8, 190-192.

77. Goleb, J.A. Anal. Chem. 1966, 38, 1059.

78. Willis, J.B. Nature (London) 1965, 207, 715.

79. Instrumentation Laboratory, Inc. "Clinical Applications of Atomic Absorption/Emission Spectrometry." Instrumentation Laboratory, Inc., Lexington, MA, 1970, pp. 23-27.

80. Marmar, J.L., Katz, S.A., Praiss, D.E., and DeBenedictis, T.J. Fert. Ster. 1975, 26, 1057-1063.

81. Iyengar, G.V., Kollmer, W.E., and Bowen, H.J.M. "The Elemental Composition of Human Tissues and Body Fluids." Verlag Chemie, Weinheim, 1978, pp. 24-27.

82. Olson, A.D., and Hamlin, W.B. Clin. Chem. 1969, 15, 438.

83. Zinterhofer, L.J.M., Jatlow, P.I., and Fappiano, A. J. Lab. Clin. Med. 1971, 78, 664.

84. Brode, K.G., and Stevens, B.J. J. Anal. Toxicol. 1977, 1, 282-285.

85. Fernandez, J.F. Chin. Chem. 1975, 21, 558.

86. Baily, P., Norval, E., Kilroe-Smith, T.A., Skikne, M.I., and Rollin, H.B. Microchemistry 1979, 24, 107-116.

87. Baily, P., and Kilroe-Smith, T.A. Anal. Chim. Acta 1975, 77, 29-36.

88. VanLoon, J.C., and Cruz, R.B. Anal. Chim. Acta 1974, 72, 231.

89. Conley, M.K., and Sotera, J.J. "An Automated Method for the Determination of Lead in Blood." Report 10, Instrumentation Laboratory, Inc., Wilmington, MA, 1979.

90. Wenlock, R.W., Buss, D.H., and Dixon, E.J. Brit. J. Nutrit. 1979, 41, 253.

91. Holden, J.M., Wolf, W.R., and Mertz, W. J. Amer. Dietic Assoc. 1979, 73, 23.

92. Kienholz, E.W., Ward, G.M., Johnson, D.E., Baxter, J., Braude, G., and Stein, G. J. Animal Sci. 1979, 48, 736.

93. Holak, W., Krinitz, B., and Williams, J.C. J. Assoc. Offic. Anal. Chem. 1972, 55, 742.

94. Kang, H.K., Harvey, P.W., Valentine, J.L., and Swendseid, M.E. Clin. Chem. 1977, 23, 1834.

95. Levinson, A.A., Nosal, M., Davidman, M., Prien, E.L., Jr., Prien, E.L., Sr., and Stevenson, R.G. Invest. Urol. 1978, 15, 270.

96. Locke, J. Anal. Chim. Acta 1979, 104, 225-231.

97. Iida, C., Uchida, T., and Kojima, I. Anal. Chim. Acta 1980, 113, 365.

98. Szpunar, C.B., Lambert, J.B., and Buikstra, J.E. Amer. J. Phys. Anthropol. 1978, 48, 199.

99. Cutress, T.W. Caries Res. 1978, 13, 73.

100. Nechary, M.W., and Sunderman, F.W. Ann. Clin. Lab. Sci. 1973, 3, 30.

101. Bernas, B. Amer. Lab. 1973, 5, 41.

102. Bernas, B. Amer. Lab. 1976, 8, 27.

103. Bertagnolli, J.F., and Katz, S.A. Internat. J. Environ. Anal. Chem. 1979, 6, 321.

104. Katz, S.A., Jenniss, S.W., Mount, T., Tout, R.E., and Chatt, A. Internat. J. Environ. Anal. Chem. 1981, 9, 209.

105. DeAntonio, S.M., Katz, S.A., Scheiner, D.M., and Wood, J.D. Anal. Proc. 1981, 18, 162.

106. Murthy, L., Menden, E., Eller, P.M., and Pickering, H.G. Anal. Biochem. 1973, 53, 365-372.

107. Julshamn, K., and Andersen, K.J. Anal. Biochem. 1979, 98, 315.

108. Hoschler, M.E., Kanabrocki. E.L., Moore, C.E., and Hattori, D.M. Appl. Spectrosc. 1973, 27, 185.

109. Gallorini, M., Orvini, E., Rolla, A., and Burdisso, M. Analyst 1981, 106, 328-334.

110. Bergman, S.C., Ritter, C.J., Zamierowski, E.F., and Cotherm, C.R. J. Environ. Qual. 1979, 8, 417.

111. Ritter, C.J., Bergman, S.C., Cotherm, C.R., and Zamierowski, E.E. Atomic Abs. Newslett. 1978, 14, 70-72.

112. Brown, G., and Newman, A.C.D. J. Soil Sci. 1973, 24, 339.

113. Van Loon, J.C., and Parissis, C.M. Analyst 1969, 94, 1057.

114. Nadkarni, R.A., and Morrison, G.H. Anal. Chim. Acta 1978, 99, 133.

115. Motuzova, G.V., and Obukhov, A.I. Biol. Nauki. 1980, 11, 87-90.

116. Jones, J.B. J. Plant. Nutr. 1981, 3, 77-92.

117. Garten, C.T., Gentry, J.B., and Sharitz, R.R.   Ecology
      1977, 58, 979.

118. Gorshkov, V.V., Orlova, L.P., and Voronkova, M.A.
      Byull. Pochu. im V.V. Dokuchaeva 1980, 24, 47.

119. Heanes, D.L.   Analyst 1981, 106, 172-181.

120. Cornelis, R., Versieck, J., Desmet, A., Mees, L., and
      Vanballenbeighe, L.   Bull. Soc. Chim. Belg. 1981, 90,
      289.

121. Wang, P.Y.   Hsi Hua Heueh 1980, 8, 249-250.

122. Lidmus, V.   Chemica Ser. 1973, 2, 159.

123. Katz, S.A.   Amer. Lab. 1972, 4, 19.

124. Sowers, L., and Katz, S.A.   Revista Latinoamer. Quim.
      1974, 4, 80.

125. Parker, C.R.   "Water Analysis by Atomic Absorption
      Spectroscopy."   Varian Techtron, Pty. Ltd., Springvale,
      Australia, 1976, p. 14.

126. Brooks, R.R., Presley, B.J., and Kaplan, I.R.   Talanta
      1967, 14, 809.

127. Meranger, J.C., Subramanian, K.S., and Chalifoux, C.
      Environ. Sci. Technol. 1979, 13, 707.

128. Meranger, J.C., Subramanian, K.S., and Chalifoux, C.
      J. Assoc. Off. Anal. Chem. 1981, 64, 44.

129. Katz, S.A.   Proc. NJ and Pennsylvania Water Pollution,
      Treatment and Testing Seminar, Cherry Hill, NJ, October
      1976, Rossnagel and Assoc., Medford, N.J., 1976.

130. Farrelly, R.O., and Pybus, V.   Clin. Chem. 1969, 15,
      566.

131. Bowen, H.M.J.   "Chemical Applications of Radiotracers."
      Methuen and Co., London, 1969, p. 30.

132. Mallory, E.C.   "Trace Inorganics in Water."   American
      Chemical Society, Washington, DC, 1976, pp. 281-295.

133. Mygaard, D.C., and Hill, S.R.   Anal. Lett. 1979, 12,
      491-499.

134. Korkish, J.   Pure Appl. Chem. 1978, 50, 371-374.

135. Barnes, R.M., and Genna, J.S. Anal. Chem. 1979, 51,
      1065-1070.

136. Burda, P., Liesser, K.H., Neitzert, V., and Rober, H.M.
      Z. Anal. Chem. 1978, 291, 273-277.

137. Hernandez-Mendez, J., Carabias-Martines, R., and
      Hernandez-Hernandez, P. Afinidad 1981, 38, 28-32.

138. Sansoni, B., and Iyengar, G.V. "Sampling and Storage of
      Biological Materials for Trace Element Analysis." Chap-
      ter 5, Elemental Analysis of Biological Materials: Cur-
      rent Problems and Techniques with Special Reference to
      Trace Elements. Tech. Rept. Ser. 197, International
      Atomic Energy Agency, Vienna, 1980.

139. Iyenger, G.V., and Sansoni, B. "Sample Preparation of
      Biological Materials for Trace Element Analysis."
      Chapter 6, Elemental Analysis of Biological Materials:
      Current Problems and Techniques with Special Reference
      to Trace Elements. Tech. Rept. Ser. 197, International
      Atomic Energy Agency, Vienna, 1980.

140. J.T. Baker Chemical Co. "Catalog 80." J.T. Baker
      Chemical Co., Phillipsberg, NJ, 1980.

141. Gelman Sciences, Inc. "Water-1 System." Golman
      Sciences, Inc., Ann Arbor, MI, 1981.

142. Millipore Corp. "Milli-Q Systems for Reagent Grade
      Water." Millipore Corp., Bedford, MA, 1974.

143. Osmonics, Inc. "Ultrapure Water Systems." Osmonics,
      Inc., Hopkins, MN, 1976.

144. Sybron/Barnstead. "High Purity Water for Laboratory
      Applications." Barnstead Co., Boston, MA, 1978.

145. ASTM. "1981 Annual Book of ASTM Standards, Part 26,
      Gaseous Fuels, Coal and Coke, Atmospheric Analysis."
      American Society for Testing Materials, Philadelphia,
      PA, 1981.

146. EM Laboratories, Inc., Elmsford, NY.

147. Galliger-Schlesinger Manufacturing Corp., Carle Place,
      NY.

148. Behne, D. J. Clin. Chem. Clin. Biochem. 1981, 19,
      115-120.

149. Christian, G.D., and Feldman, F.J.   "Atomic Absorption Spectrometry."   John Wiley and Sons, NY, 1970, pp. 188-195.

150. Watling, H.R., and Wardale, I.M.   In "Comparison of Wet and Dry Ashing for the Analysis of Biological Materials by Atomic Absorption Spectrometry, L.R.P. Butler, Ed. Conference Proceedings Analysis of Biological Materials, Pergamon, Oxford, 1979, pp. 69-80.

151. Hislop, J.S., and Williams, D.R.   "The Use of Gamma Activation To Study the Behaviour of Certain Metals, in Particular Lead, During the Ashing of Bone."   Proceedings of the Symposium on Nuclear Activation Techniques in the Life Sciences, Bled, April 1972, International Atomic Energy Agency, Vienna, 1972, pp. 51-62.

152. Behne, V.D., and Brumeätatter, P.   "Probleme bei der Spurenelement-analyse in der Medizin."   Spurenelemente Symposium, Bad Kissinger, 1977, Thieme Verlag, Stutgart, 1979.

153. Rice, T.D.   Anal. Chim. Acta 1977, 91, 221-228.

154. Artiola-Fortuny, J., and Fuller, W.H.   Arizona Agricultural Experimental Station Paper No. 2671, University of Arizona, Tucson, AZ, 1979.

155. Artiola-Fortuny, J., and Fuller, W.H.   Compost Sci./Land Util. 1980, May/June, 30-34.

156. Delfino, J.J., and Enderson, R.E.   Water and Sewage Works 1978, 125, R32-R34, R47-R48.

157. Feinberg, M., and Ducauze, C.   Anal. Chem. 1980, 52, 207-209.

158. Feinberg, M., and Ducauze, C.   Bull. Soc. Chim. France. 1978, 419-425.

159. Ross, W., and Umland, F.   Talanta 1979, 26, 727-732.

160. Watson, M.E.   Communic. Soil Sci. Plant Anal. 1981, 12, 601-617.

161. McKenzie, T.   "The Analysis of Trace Metals in Air by Furnace Atomization."   No. AA-7, Varian Techtron, Pty, Ltd., Mulgrave, Australia, 1980.

162. Brown, E., Skougstad, M.W., and Fishman, M.J. "Methods for Collection and Analysis of Water Samples for Dissolved Minerals and Gases." U.S. Geological Survey Techniques of Water Resources Inventory, Book 5, Ch. A 1, U.S. Geological Survey, Washington, DC, 1975.

163. Fishman, M.J., and Brown, E. "Selected Methods of the U.S. Geological Survey for the Analysis of Wastewaters." Open File Report. 76-177, U.S. Geological Survey, Washington, DC, 1976.

164. Horowitz, W., Ed. "Official Methods of Analysis," 12th ed. Association of Official Analytical Chemists, Washington, DC, 1975.

165. "American National Standard on Photographic Processing Effluents." American National Standards Institute, New York, NY, 1975.

166. Adelman, H, Jenniss, S.W., and Katz, S.A. Amer. Lab. 1981, 13, 31-35.

167. Goyer, R.A., and Mehlman, M.A. "Toxicology of Trace Elements." John Wiley and Sons, New York, NY, 1977.

168. Henry, R.J., Cannon, ᴐ.C., and Winkelman, J.W. "Clinical Chemistry Principles and Techniques." Harper and Row, Hagerstown, MD, 1974.

169. Tietz, N.W., Ed. "Fundamentals of Clinical Chemistry." W.B. Saunders, Philadelphia, PA, 1976.

Index

Index

275